Inquiring About Plants:
A Practical Guide to Engaging Science Practices

Inquiring About Plants

A Practical Guide to Engaging Science Practices

Gordon E. Uno

Marshall D. Sundberg

Claire A. Hemingway

Botanical Society of America | *St. Louis, Missouri*

ISBN-13: 978-0-9848582-2-4 (paperback)
ISBN-13: 978-0-9848582-3-1 (epub)
ISBN-13: 978-0-9848582-4-8 (mobi)

Design and composition by Kristina Kachele Design, llc
Prepress by iocolor, Seattle
Printing and binding by Thomson-Shore, Inc. Dexter, Michigan

Cover images: Nicole Hughes and Simon Uribe-Convers; back image, Seana Walsh. The credits section for this book is considered an extension of the copyright page.

This material is based upon work supported by the National Science Foundation under Grant No. 0733280. Any opinions, findings, and conclusions or recommendations expressed in this material are those of the author(s) and do not necessarily reflect the views of the National Science Foundation.

Second printing, 2013

Contents

Preface

Science is an endeavor of asking and answering questions about the natural world and how it works. The practice of doing science does not occur in a vacuum—biologists working in a lab or in the field seek to understand a particular research area in a particular context. For science learners in the classroom, teachers play pivotal roles in providing the content and context for their students.

This is a book about helping students engage in the practices to ask and answer scientific questions. That means this is also a book about assisting teachers to create an engaging environment for deep learning. While the approach modeled here is relevant to any scientific discipline, we use botanical examples that we have found successful with students in the classroom.

Our purpose is to make inquiry teaching and learning about plants more accessible. Neither a traditional textbook nor instructional manual, this practical guide combines biology content, instructional strategies, and learning research. Diverse examples of plant activities encompass ecology, evolution, structure and function. Our aim is to provide a concrete view of what it looks like to integrate science content and science process. The teaching notes associated with activities include suggested probing questions to help students along the way to carefully consider the nature of the evidence and what you might expect as typical student responses. We do not intend that teachers use the scripts verbatim. Rather this structure is intended as a flexible framework for teachers in working with their students.

All three of us have botany as part of our scientific expertise, and because this project emerged from a grant related to the PlantingScience project, most of the examples in this book are about plants and their close relatives. The teaching and learning methods included here, however, are applicable to any group of organisms or biological concept. The focus on inquiry is a natural extension of the work in which we have been engaged throughout our professional careers.

We have written this book for secondary science teachers, as well as college instructors. For some practitioners, this book may answer how to create a culture of student-centered inquiry, and our aim is to offer classroom-tested approaches if you are looking for ways to change how you teach. For those of you who have already embraced the power of inquiry in your classroom, we provide diverse ideas to enrich what you already do.

The book begins with the question, "What do you want your students to know, value, and be able to do by the end of your class?" The journey to address this overarching question is organized in three sections.

Chapter 1 takes a deep look at what you want your students to value and be able to do. This section addresses major critical thinking and science practice skills: making careful observations, asking questions based on observations, developing hypotheses and making predictions, designing investigations, collecting and analyzing data, working with representations, using mathematics, working with scientific evidence, communicating results and conclusions, and relating content knowledge to other concepts. Each of these is illustrated with one or more detailed plant activities. Across these activities, you will see several themes begin to emerge. One is that we are working towards having students actively generate their understandings of biological concepts throughout the course. Another is the role of discussion and reflective questioning—within pairs, teams, or across the entire class—to support that process. And a common topic we emphasize in discussions is considering alternative explanations when working with scientific evidence.

Chapter 2 deals with what you want your students to know. This section dealing with content tackles the issues of how much and what content to select for your course and ways to help you and your students think about thinking. It also provides evidence of the effectiveness of these teaching and learning approaches. It then moves on to focus on two strategies that work together to support students as they construct their understandings: focusing on the big ideas of biology and dealing with misconceptions. Like the previous section, these are illustrated with plant inquiry activities.

Chapter 3 considers how to build on these foundations to create opportunities for your students to connect and apply their knowledge of biological concepts. Evolution is used as an example for weaving a theme throughout your course, rather than covering it as a separate topic. Climate change is used as an example for integrating inter-disciplinary evidence to address current issues facing society. Finally, the book concludes with a focus on students as independent investigators. Through the repeated opportunities to integrate content and process skills throughout a course, students will build their experience working as scientists. No matter what career paths students take in life, we believe our role is to help students come away from a biology experience with a stronger capacity and curiosity to ask and answer scientific questions.

This book developed from our collaborations in the PlantingScience outreach program, which supports student-centered investigations by making scientists accessible to classrooms as online mentors. At the end of the book are a number of references and resources that you may find helpful as accessory materials. Within the book we mention only two of the many plant investigation resources that are available online at www.PlantingScience.org; please see the website for additional resources including techniques for investigations and links to education materials by program partners.

Acknowledgments

This book represents the ideas and activities developed and collected over many decades of work. We appreciate all the efforts of our colleagues who have contributed to the information in this book. We especially thank Catrina Adams, Beverly Brown, and Simon Malcomber for reading various drafts of this material.

We are indebted to the many teachers, students, and mentors in Planting-Science that have made this project possible. The teachers with whom we have worked during summer workshops and in field-testing materials have been a source of invaluable insights and inspiration. It is an ever-exciting thrill to see where discovery can lead students when they have authentic opportunities to engage in the scientific endeavor.

The images on the cover and chapter title pages were submissions to the Botanical Society of America's Triarch Botanical Images Student Travel Award. We thank the individuals for use of their photographs. We also thank the Botanical Society of America for its commitment to botanical education and outreach.

We gratefully acknowledge funding for this project from the National Science Foundation (DRL-0733280). Any opinions, findings and conclusions or recommendations expressed in this material are those of the author(s) and do not necessarily reflect the views of the National Science Foundation (NSF). Monsanto Fund provided start-up funds for curriculum development.

Gordon E. Uno, University of Oklahoma
Marshall D. Sundberg, Emporia State University
Claire A. Hemingway, Botanical Society of America

1

Incorporating Science Practices In The Classroom

Before you begin developing your course, or any part of it, you first need to ask, "by the end of my course, what do I want my students to **know, value, and be able to do**?" The "know" part is content, but let's don't worry about that first. The "value" part has to do with your students' interests and feelings about biology. For example, do you want your students to like biology as a subject? If so, what are you going to do during the class that will help them appreciate the wonder of living things and the process of science? Perhaps you could incorporate investigations outdoors, which might increase students' appreciation of nature. You cannot assume that, just because biology is interesting to you, it will automatically be interesting to students. You have to be "intentional" about building something into your course that will help students discover for themselves how interesting living things are.

The "do" part of student outcomes is where you should concentrate most of your thinking as you develop your course. This part involves critical thinking and science process skills that you want students to possess and to be able to use competently by the end of their experience in your class. For instance, is it important that your students be able to design an experiment (it should be!), and to work with data (such as on graphs and tables), or to make careful observations and ask good questions that could lead to independent investigations? These are skills that will be incredibly valuable to students as they move through their academic and professional lives—whether they major in science or not. Again, however, if you want your students to be able to work with datasets (for example), then you have to give them practice with data during your course—you must be intentional. If you want them to be able to design an experiment—you have to be intentional in your course outline to provide the opportunities (plural) for students to practice their thinking and science process skills as related to designing experiments.

The following list includes the most important science process skills that your students should possess. As you will see, they are all important components of a scientific method of investigation—a scientific method. Remember, however, that scientists rarely follow the same order of these components every time they design an experiment or investigate a scientific question. These components, however, are key to how you should consider designing your course. You should think about getting your students to practice at least one of these skills EVERY CLASS PERIOD. It would be difficult to practice all of them on any given day, however, you should engage your students in some thinking or process skill every single day.

Critical Thinking Skills and Science Practices

1) Making Careful Observations
2) Asking Questions Based on Observations
3) Developing Hypotheses and Making Predictions
4) Designing an Experiment (or Observational Study) to Investigate Natural Phenomena
5) Collecting and Analyzing Data
6) Working with Representations (Graphs, Data Tables, etc.)
7) Using Mathematics Appropriately Throughout a Class
8) Working with Scientific Evidence and Alternative Explanations
9) Communicating Results and Conclusions Based on Evidence
10) Connecting and Relating Content Knowledge to Other Concepts to Construct a Framework of Understanding Biology

This book includes several "tricks of the trade" that you can use in your class to promote critical thinking and science process skills while teaching the content you want. In terms of content, remember that there is so much content in biology that no teacher at any level should try to cover all of it in any class. Focus first on major concepts, and then only on those details that are critical to the understanding of those concepts. This book will also provide some tips about how to approach the selection and teaching of biological content.

This section includes several examples for each main category. They range from a short, simple experience focused around one idea to a sequence of related steps that build multiple skills simultaneously and introduce several biological concepts. The longer activities contain background information as well as suggested instructional cues to guide your students through the sequence. To illustrate how one simple experience can be extended as far as you'd like for your students, activities using the fence photograph are carried through making observations, generating questions, forming predictions and designing experiments. One aim of this book is to provide diverse examples of botanical content in context. So, the activities cover ecology, anatomy, physiology, diversity, evolution, and much more.

Photographs of Nature Spark Observations, Questions, and Experiments

Use any photograph that illustrates a seeming biological anomaly or mystery, or some biological principle, to generate a discussion about a scientific method of discovery (remember, there is no such thing as THE scientific method—scientists investigate the world and begin their studies in many different ways).

Pictures of ecological phenomena are particularly good for this activity. For instance, the bare zone under a tree can be used, or this picture of an open prairie with a fence running down the middle of it.

Ask your students, **"What do you see?"** The key is to keep asking questions, even after you get the "right" answer. It is equally important for students NOT to answer with the "cause" of the phenomenon—you want just what they can see. For instance, in this photograph, students may say, "I see that the field on the right hand side was grazed." This is a conclusion or speculation because there are no grazing animals in view, thus it is not an appropriate answer. . . . at this time. Students can't see any grazing animals, so it's not an appropriate observation. Of course there is a lot of cow manure in the field on the right, but it may also be on the left....just harder to see. So, what they CAN observe is a grassland area with taller grasses on the left side of the fence and much shorter grass and cow manure on the right. Of course, there are many other observations that students may make. Just keep asking, and make positive comments about every observation. Also, remember what you were taught about wait time! List all observations, so all students can see them.

Now that you have a set of **observations**, ask students if they have any questions that they could ask about those observations—what might be the cause of the differences that they have noted or the observations that they have made? When they come up with a **question**, ask them what was the evidence (the observations) that triggered that particular question. Now is the time that they can ask if grazing caused the difference between the fields on the left and right sides of the fence. A student might say that the observed differences were caused by animals grazing on the right side of the fence and not the left because the grasses are shorter on the right. This speculation is based on what appears to be cow manure on the right, which could have been left by the grazing animals. Again, students might have other questions related to cause of the short grasses (or taller grasses on the left---perhaps the left side was fertilized)....again list all questions, but make sure that students have linked their questions to what was observed.

Now that you have a set of questions, choose one of them to test by getting students to **design an experiment**. While you can use any of your students' questions, you might start with the "grazing" question. One reason is that those students who "know" that grazing caused the difference have a much harder time designing an experiment to support their speculation. That is why it is so important to keep asking for more observations and questions even after some know-it-all student blurts out "grazing caused the difference!" By doing so, you can bring more students into the conversation, and you demonstrate that it's not about who can say the right answer, but who can figure out how we can investigate the situation—how do we know something?

So now, if we suspect that grazing caused the observed differences, then how might we investigate that to determine if it is "true?" One experiment might be to put a barricade surrounding a small area (an exclosure) inside the field on the right hand side of the fence. Then ask the question "If grazing is responsible, what do you **predict** will happen to the area inside the exclosure over time?" What do students predict will happen both inside and outside the exclosure? One might predict that grasses will grow again inside the exclosure while they will remain small outside the fenced area due to continued grazing outside the exclosure, but not inside. Of course, both inside and outside the exclosure may remain in the same condition or change in the same way if all of the grazing animals have already been removed from the field. Or...one could put cattle in the field on the left and not in the right side. One would then predict that the grasses would be eaten on the left, and the cow pies would pile up on the left hand side, resembling what had already occurred in the right-hand side. In all cases, remember to ask students what they **predict** will happen if their testable explanation (hypothesis) for the cause of the difference is "true" (supported by evidence gained through their experiment).

An important point here is that it is not essential that you carry out the experiment that is designed—that would be desirable, of course, but may not be possible. The process of making careful observations about some biological phenomenon, asking good questions based on those observations, and then designing an experiment based on a question (and predicting what may happen) are all desirable skills to practice, practice, practice throughout your course.

1) MAKING CAREFUL OBSERVATIONS

Similarities in Nature

In a nutshell: The point here is that students often think objects appear to be similar to each other. However, if they look closely (make careful observations), they will see that the objects really are different from one another.

Divide students into teams before going outside. Ask each team to "find four (4) natural (not man-made) objects that appear identical to them." For example, pine cones or rocks or leaves of similar size, shape, and color—the more similar the better for this activity! Students return to class with the four objects, and label them "A, B, C, and D." You can use a small piece of masking tape for each object. Then, each team, as a group, writes a description of each of the four objects—the descriptions need to be written clearly enough so that someone who has not yet seen the objects can tell which object is which based on the descriptions. Then, the descriptions (placed in random order) and the objects in random order are given to a different team. Although the four objects are labeled (A, B, C, or D), the descriptions are not. The second team uses the descriptions to figure out which object is which.

Teams can trade objects and descriptions several times. Each team needs to receive "feedback" from the other teams to determine how well written all the descriptions were: e.g., could they detect differences between the objects based on their careful observations? If few other teams are able to identify the correct objects based on the descriptions, either the first team (that made the descriptions) did not observe carefully enough or write quality descriptions to help the other team(s) distinguish objects.

Descriptions of Tree Bark

In a nutshell: A benefit of this sequence of observations is the apparent simplicity of the task, which will reinforce students' naive expectations that a superficial answer is all that is required. Sharing will describe some unexpected complexities that encourage students to be more critical upon their second observation—learning point to be stressed the rest of the course.

This works best with a small grove of trees available so each student can concentrate on the bark of her/his own tree. (A small, 1ft^2, patch of lawn can be substituted.) The students' task is to spend five minutes making notes and observations about the bark of "their" tree to be able to describe it to others. Their challenge is to be the one to find and list the most descriptive characteristics. Students then share their individual observations with the class and are questioned by their peers. Depending on class size, this may take 15–30 minutes.

Figure 1.1 Some examples of tree bark. A. American Elm. B. Sycamore. C. Red Cedar. D. Hackberry.

Most student observations will initially be superficial—the bark is smooth or rough, color of the bark, the presence of moss or lichens, etc. However, some individuals will come up with more sophisticated observations such as distinctive layers of bark, insects or evidence of insects, fungi, differences between the surface and the "cracks," sap or other exudates, etc. After sharing, each student is given 10 more minutes back at their tree to reexamine their bark and add to their original list.

One follow-up assignment could be to write a one- or two-page paper describing their observations of the tree bark. Another possible activity is to begin a discussion of what might be a function of a particular observed character. Ask students, "Can you make a hypothesis about the function? If so, how could you design an experiment to test your hypothesis?"

This activity is based on one originally described by Professor W.J. Beal in 1880 at what is now Michigan State University. He suggested that, in winter, tree branches about 2 feet long could be brought into class for observation.

Interpreting Microscopic Images

In a nutshell: The following two activities illustrate that, when you are interpreting what you see in the microscope, you have to consider how the slide was made and if the structure you are seeing is real or an artifact of preparation. Have students work in teams to discuss the images provided. Then guide a class discussion to encourage students to describe their observations and evaluate what they see.

In the following, suggested teacher questions are in roman, *possible student answers are in italic type*, and background information to interpret the questions are in roman sans.

ELECTRON MICROSCOPE IMAGES

⇢ The image shown is a collage of three separate photomicrographs as seen with an electron microscope. How many different kinds of cells do you think these images represent and what evidence do you have for your answer?

   These are three images of the same cell type, but prepared in different ways.

⇢ What evidence do you have that the left and middle images are plant cells?

   A cell wall is evident around the boundary of cells.

⇢ Describe the cells in the left-hand image.

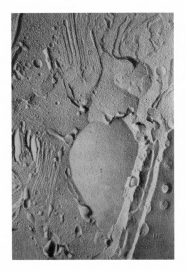
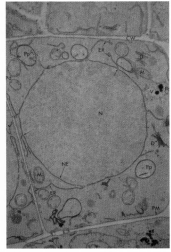

The cell walls are dark and the cytoplasm is packed with dark granular structures either organized into organelles or scattered throughout. In the cell on the upper left there appears to be a dark nucleus surrounded by a larger light circular structure.

Figure 1.2 How many different kinds of cells are in these three electron microscope images? (From Ledbetter and Porter, 1970)

This cell is cut very thin and stained particularly for nucleotides and carbohydrates. The adjacent cell walls are dark as are ribosomes in the cytoplasm and the nucleus. In the nucleus, spread out DNA is evident in the lightly stained large circular region of the upper left cell. RNA in the nucleolus is the darkly stained round structure within the nucleus.

⇢ Describe the middle cell. How large is it compared to the left-hand cell? What is your evidence?

The outlines of organelles are clearly visible in this image but the cytoplasm does not appear to be as darkly stained as in the previous slide. This cell is about the same size as the previous one. The outline and thickness of the cell walls are about the same and the nucleus appears to be about the same size.

This cell was also cut very thin, but this time it was stained particularly for membranes. The cellulose cell walls do not pick up any stain. The thickness of the cell walls, and the outline of individual cells appear to be the same size.

↦ Why do cells appear to be outlined in black?

The cell membrane on each side of the shared cell walls is stained.

↦ In each cell, the cell membrane is pressed against the wall so in the image the shared cell walls appear to be outlined. In several places there are lines connecting the cells across the shared cell wall between them. What does this suggest?

The cytoplasm of individual plant cells is connected.

In most plant cells there are tiny holes through adjacent walls so the cell membrane from one cell can connect directly to the cell membrane of the next cell.

↦ The large round structure in the center of the cell is the nucleus. Describe the structure of the nuclear membrane as you go around the periphery.

In some places the nuclear membrane is attached to other membranes and the membranes of other organelles. In some places there appear to be gaps in the nuclear membrane.

Although not visible here, the nucleus has a double membrane with pores in some places. The outer membrane can connect to other organelles through the network of endoplasmic reticulum.

↦ Although these cells look very different, all three actually came from the same tissue. This is a good reminder that when you are interpreting what you see in the microscope, you have to consider how the slide was made. Are you seeing an accurate representation of what the cell looked like when it was alive or are you seeing an artifact of preparation?

The appearance of cells, killed and stained for microscopy, depends on how they were killed, and what procedures were used to prepare the slide. The right hand cells were not even cut. Instead, the tissue was frozen and then cracked. Just like in ice, cracks tend to form in weak spots where particles are located. In this case the particles were membranes so the cracks expose the inside and outside if different organelles. This image is at higher magnification than the previous two. On the lower right are the double membranes of the nucleus with tiny pores. To the left and above are two stacks of Golgi membranes with round vesicles budding off the ends of the stacks.

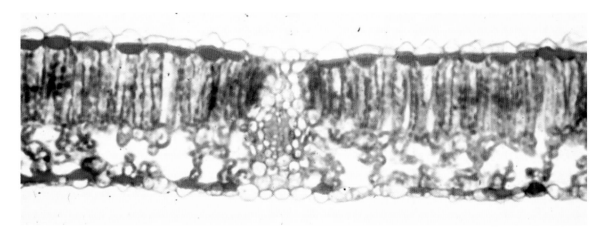

→ Above is a photomicrograph of a cross-section through a dicot leaf. The outer layer of cells around the entire leaf is the epidermis. Describe the appearance of the epidermal layers. Is the epidermis one or two cells thick? How do you know?

Figure 1.3 Photomicrograph of a cross-section through a dicot leaf.

In the upper epidermis there is a clear layer on top and a red layer below. The same is true on at least some of the lower epidermis. Most students will interpret this as two cell layers; other students might suggest that it is a single layer of cells.

This slide is of a maple tree that had purplish colored leaves. The vacuoles of most epidermal cells were filled with a red pigment that also picks up the red stain used to prepare this slide. When making microscope slides, the cells in the tissue are usually quickly killed with a solution containing alcohol among other chemicals. Then the cells are dehydrated with alcohol solutions of increasing concentration to remove all the water. This tissue was dehydrated too quickly and as the vacuoles of the epidermal cells shrank, the cytoplasm pulled away from the outer walls. This is called plasmolysis. What appears to be two layers of cells, an inner red layer and an outer clear layer, is a single layer of plasmo-lyzed cells. Again, this is a good reminder that when you are interpreting what you see in the microscope, you have to consider how the slide was made and if the structure you are seeing is real or an artifact of preparation.

Ten Discrepancies

In a nutshell: The advantages of this activity are, first, that it reinforces good observational skills, and second, that it stimulates design of meaningful research projects to address original questions.

This activity originally was designed as a team project to be done as a homework assignment out-of-class in preparation for choosing a research question to examine. The assignment is to meet as a group at some outside location, such as a park, a zoo, or other natural area, and spend enough time walking through the area and observing to be able to write down at least ten things that seem to be unusual compared to everything else. For instance, a leaning tree on the side of a river bank while other trees of the same kind may be upright, branches may extend further out on only one side of a tree, moss or lichens may be found mostly on one side of a tree, squirrels may build nests in certain trees but not others, etc.

Individual students should bring to class a list of their ten discrepancies with at least one question about each one. In addition to bringing a list, asking that students document these via hand-drawn illustrations or photographs could reinforce the scientific habit of recording observations. In class the group should decide on one question to examine experimentally, make a testable hypothesis or prediction about that question and design an experiment to test that hypothesis about what may be the cause of a difference or discrepancy that the students found.

Movements in Time Lapse

In a nutshell: The benefit of this activity is that time-lapse photography reveals processes in a plant's life that take place on a time scale that humans can't see first hand. New ways for students to observe plants can elicit new questions.

Select a series of time-lapse movies on plant movements, such as those created by Roger Hangarter available on Plants-in-Motion:

http://plantsinmotion.bio.indiana.edu/plantmotion/movements/tropism/tropisms.html
http://plantsinmotion.bio.indiana.edu/plantmotion/movements/nastic/nastic.html

http://plantsinmotion.bio.indiana.edu/plantmotion/movements/
leafmovements/clocks.html

This works best when the movies on tropic responses, nastic movements, and circadian movements are shown without the type of movement being identified. The point is for the movies to stimulate students to wonder about the external and internal stimuli responsible for the plant movements and to think about possible advantages to the plants that move. For each time-lapse movie you show to the class, ask students to individually note what they think is happening in the movie and one question it raises for them about plant movements. Then have students work in pairs to discuss their questions about what the plants might be responding to and the evidence from the movie.

This activity could be followed up with the Darwin and Phototropism activity, where students consider the evidence and alternative explanations of Charles and Francis Darwin's experiments on phototropism. Students could also make their own time-lapse videos with plants in your classroom. There are many tips on the Internet about how to make time-lapse movies using a cell phone or digital camera. Student teams could be responsible for creating their own time-lapse movie of plants moving, either as a way to further generate and explore questions or at the end of the lesson as a final product to bring together their ideas on plant tropisms.

Another approach to make plant movements more noticeable to students is with a subtle classroom demonstration. Without announcing or drawing attention to your action, take one of the potted plants in your classroom and place it on its side. Seedlings or small herbaceous plants work well for this. Depending on the plant you use, bending of the plant stems should be evident within several hours or by the next class day.

3) DEVELOPING HYPOTHESES AND MAKING PREDICTIONS

Fruit and Seed Sorting

In a nutshell: The purpose of this activity is for students to develop, explain, and test their rules for grouping objects. Sometimes the object is placed into the "correct" group, but for the wrong reason. This is similar to when a hypothesis is tested and the results support the hypothesis, however, the hypothesis itself is incorrect.

Students are separated into teams (2–5 students). Each team is given a set of dried seeds and fruit that share some similarities but that are not identical to each other. The set could be dried goods from the kitchen pantry such as legumes, castor beans, corn or a variety of spices such as whole pepper, allspice, whole cloves, cardamom, mustard, fenugreek, etc. Include a variety with similarities and differences in size, shape, and color within the set of 10–15 objects. You could also intentionally choose some that are commercially called seeds but are botanically categorized as fruit (e.g., black peppercorns are dried berries; whole coriander and fennel are dry indehiscent schizocarps).

The directions to each team are to separate the objects into as many groups as the team would like. However, all members of the team have to agree on the grouping and why the figures are grouped together. There should be at least 2 different groups (i.e., the team can't say that there is just one group of dried seeds). Next, the team makes a list of the groups (using the letters) and then that list is given to a different team. The second team must try to figure out why the first team grouped the objects the way they did—this is the hypothesis about what Team A did and why (e.g., were all of the objects round and blackish). The second team tests this hypothesis by placing a new object (not seen before by either team) into the group of objects they think the first team would. Then the two teams talk to each other and see if the reasons for grouping are correct and if the placement of the object is correct.

Convergent Evolution in Succulent Plants

In a nutshell: This activity builds on earlier activities of observing individual plants and patterns in nature. Students use observations about the shape and external features of a collection of organisms to develop ideas about what similarities and differences in form might mean. The activity introduces form-function relationships as well as evolutionary relationships. Many plants adapted to desert conditions have evolved succulent morphology to store water. That is, succulent organs have enlarged cells in their ground tissue (parenchyma cells) with huge central vacuoles that store water. Roots, leaves, or stems may exhibit succulence, and similar adaptations have evolved many times in several plant groups.

Each of the image panels below contains three individuals, two of which are more closely related than the third. The students' task in each case is to predict which two are more closely related and have a reason for their prediction.

Students will usually link A and C in Figure 1.4 because of the green color.

In this case, A and C are more closely related because they are both plants: A is an African euphorb and C is an American cactus. B is an animal, an echinoderm commonly called a sea urchin.

A   C

B

Figure 1.4
Predicting relation-
ships among three
organisms — I.

A   C

B

Figure 1.5
Predicting relation-
ships among three
organisms — II.

Typically students will place A and C of Figure 1.5 together because of the pin-cushion shape.

A and C share the same pincushion shape, but A and B are cacti from the Americas. C is an African euphorb. The family Cactaceae is native to North and South America (with the exception of the epiphyte *Rhipsalis,* which also occurs in Africa, Madagascar, and Sri Lanka. Competing hypotheses for how it arrived in the Old World include introduction by sailors traveling along trading routes, dispersal by migratory birds, or being a Gondwanan relict—but that is a fascinating story for another day).

Any cactus-like plant from Africa, Asia, or Australia belongs to a different family because of their evolutionary history. All three species have ridges and valleys that function like pleats. After a rain, the tissues swell and the pleats expand. During drought, the tissues shrink and the ridges produce shade. The shaded side of a ridge can be several degrees centigrade cooler than the sunny side.

Figure 1.6 Predicting relationships among three organisms — III.

Again, students will typically lump A and C together because of the similar shape.

These three are stem succulents like the previous set. Again B is an elongate, upright form while A and C are only slightly elongate. They all have ridges and valleys. A and B are African euphorbs and C is the American cactus. If there are no flowers, the easiest way to tell a cactus from other stem succulents is to look at the spines. Cacti always have clusters of spines, as in C, because they are modified branch shoots that form in the axils of ephemeral leaves that last only a week or two following a rain. Each leaf of the axillary branch shoot develops as a spine. Euphorbs have only one or two spines associated with the leaf scar.

Recognizing a Workable Hypothesis

In a nutshell: The aim here is for students to see that posing hypotheses and predictions is about working out testable scientific explanations—not about "getting the right answer", as you have likely heard them comment when an experiment ends. Another advantage of getting students to consider alternative hypotheses is that it gives them practice reasoning and evaluating ideas they can use later when interpreting data and drawing conclusions. The first example using redwoods drives home that multiple factors often play a role in any biological phenomenon and scientists seek to understand which influences are strongest in particular conditions. The second example illustrates a sequence to move from very broad explanations for a pattern in nature to get to a testable relationship.

Figure 1.7 Coastal Redwood trees growing in two different locations. A. *Sequoia sempervirens* trunks at Redwoods National and State Park, California. B. 55 year-old *Sequoia sempervirens* bonsai tree from Brooklyn Botanic Garden, New York.

Have students work in pairs to view and discuss the images of coast redwoods (*Sequoia sempervirens*). In their natural environment, these trees are record breakers for plant height with many reaching 300-350 ft. Yet in bonsai form, coast redwoods can sit on a table top as an ornament only a couple of feet tall.

First, have student teams describe what they notice by comparing the redwoods in these photographs. Even without scale bars in the photos, the dramatic size differences between individuals of the same species will stand out to students. As teams report out, their observations can be used to describe a relationship in nature about plant growth as a response to environmental conditions.

This statement of the relationship in nature of a plant with its environment could then have several possible explanations (hypotheses).

To lead students toward considering alternative causes for the relationship that could be investigated, then ask students, "What environmental factors influence plant growth?"

List all student ideas about factors that promote or limit growth as they are offered. Students will likely initially offer elements generally important: sunshine and water critical to photosynthetic processes converting carbon dioxide and water into the sugars and starches that fuel plant growth. From everyday experiences with houseplants or gardens, students may suggest mineral nutrients found in the soil or soil makeup (pH, soil texture). The small size of the bonsai container will likely raise ideas about the relationship between the amount of soil or space available and plant size. Also, ask students if a giant plant such as a redwood has any limiting factors to its growth (which it does, of course, and that is the reason why the world isn't covered with redwood trees).

Predicting an outcome for an investigation is often the easy part for students. The tricky part is connecting the expected outcome to the explanation being tested. Have student teams compare examples that DO NOT and DO include statements of why or under what conditions the outcome is expected. While using your own previous students' work can be very effective, here is an example that you can use to get started.

> "If a plant doesn't receive nutrients from the soil, then it will grow smaller than a plant that receives nutrients."
>
> "If nutrients in the soil are a primary limiting factor for plant growth, and I don't give all plants access to these nutrients, then plants grown without nutrients will be smaller than plants receiving the nutrients."

After students compare wording for the above hypothesis about the role of soil nutrients on growth, have student teams pose an alternative hypothesis and prediction. Then student teams swap papers with another team, and students attempt to identify the explanation that the fellow team's hypothesis is testing. Give all students an opportunity to re-phrase any statements that lacked obvious reasoning about why the outcome is predicted. As teams report out their refined alternative hypotheses, use the list generated to lead into a class discussion about which would be more important to test, and which more practical to test.

It is certainly possible for students to experimentally test in a classroom setting some hypotheses generated from their observations of the coast redwood photographs using plant species that germinate and grow relatively quickly. However, it works well to consider alternative hypotheses as thought experiments, without necessarily going on to design and conduct investigations.

Another strategy: This second example is a thought exercise using plant hairs—a topic not well known by students or experimentally tested by scientists. The idea that plants have hairs on their leaves, stems, and flowers (in addition to root hairs) might be novel and intriguing to students. An advantage of using something unfamiliar to students is that it can free them from seeking to pull the "right answer" from their memory banks of memorized information from earlier classes. That might make students more open to offering and considering alternative explanations for biological phenomena.

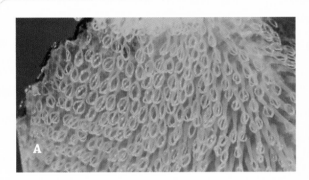 

Figure 1.8 Examples of two types of plant hairs. A. Egg beater hairs on the leaves of *Salvinia molesta*. B. Autofluorescence of simple hairs on the petiole of *Coleus*.

Have students examine a diversity of plant hairs (trichomes, in the botanical lexicon). In an ideal world, every teacher would have access to a greenhouse stocked with a many kinds of plant species to illustrate plant diversity and morphological characters. If this is not the case, you could provide a series of images of plant hairs or guide students on a web quest to identify 10 plant species that have hairs that look different in some way (in an unrestricted web quest, be prepared to talk about *Cannabis* with your students).

Have students make notes about their observations, such as where the plant hairs are located on the plant (leaves, stems, or other structures) and their overall look (glandular and sticky or not).

Discussing the array of plant hair types raises the general question of why do plants have them. Why questions are somewhat philosophical in nature and therefore not testable. So instead ask students, "What are some possible functions of plant hairs?" As students offer their ideas, make sure they describe the observations and plant names that they are thinking about. Depending on the plant species students examine, they may offer possible explanations that hairs function to protect plants from herbivores, pathogens, extensive light, extreme temperatures, or function to absorb water, or secrete salt, or attract animals (all of these, and more, are functions of hairs in various plants!).

From the list of alternative hypotheses and plant examples generated, students will likely grasp that explanations of what plant hairs do in nature varies a great deal and depends on the species they are interested in investigating. Thus, a broad hypothesis about the function of hairs in plants (there are some 400,000 known species on Earth) is not really a workable hypothesis. But it is possible to narrow the scope and generate a testable hypothesis about plant hairs for particular species.

If students have not previously mentioned hairs of carnivorous plants, this in an excellent opportunity to take advantage of students' curiosity about carnivorous plants in the context of considering alternative hypotheses.

Sundews, members of the genus *Drosera*, have sticky glandular hairs. Show students images of an insect caught in the sundew hairs (e.g., http://www.botany.org/carnivorous_plants/Drosera.php). Students can connect the observations that sundews can trap insects to their ideas of possible functions for this trait generated during the earlier discussion. Have students pose alternative hypotheses about possible functions of *Drosera* hairs. From the list generated earlier, students will likely identify that hairs could serve a protective function to stop herbivorous insects from damaging the sundew or hairs could serve to attract insects so that the sundew can extract their nutrients. Have students generate expected outcomes (predictions) of experiments based on these alternative hypotheses.

After students pose alternative hypotheses and share each other's work to identify the explanation being tested, students could compare their ideas to Charles Darwin's experiment. The ability of *Drosera* to catch insects so captured the curiosity of Charles Darwin that he ran many experiments on *Drosera* and other carnivorous plants. Darwin reasoned that if the sticky hairs serve a nutritional function, and, if he fed some *Drosera* insects and stopped others from trapping insects, then the plants denied insects would be less healthy. Was Darwin's original hypothesis supported or not?

Resources:
World Record Plant Height: http://www.npr.org/blogs/krulwich/2011/04/08/135206497/the-worlds-tallest-tree-is-hiding-somewhere-in-california).

Darwin Online – The Digestive Power of the Secretion of *Drosera*: http://darwin-online.org.uk/content/frameset?pageseq=11&itemID=F1217&viewtype=side

Attack of the Killer Tomatoes, popular article on plant hairs: http://www.independent.co.uk/news/science/attack-of-the-killer-tomatoes-1834638.html

Sticky-fingered Plant May Hold the Secret to Snaring Bed Bugs, popular article on research: http://www.wired.com/wiredscience/2013/04/biomimetic-bedbug-snare/

Corn Competition

In a nutshell: This activity is a fun, low-stakes way for students to learn about the importance of a control in interpreting the results of an experiment. It works well as a preliminary assignment early in the year to prepare students for the opportunity to design their own experiments.

This activity is presented to students as a contest to "grow the tallest corn plants," not as an opportunity to design an experiment. Students have three weeks to grow plants that are taller than plants grown by other students in the class. A key piece in setting up the contest is to give students a choice: would they like to receive one or two pots in which to grow their plants? All students will receive vermiculite (to use as a potting medium) and five corn kernels for each pot (a total of either 5 seeds if they choose one pot or 10 seeds if they choose two pots). Starting with these materials, student can grow corn plants using any conditions of their choosing.

Students are tasked with bringing plants labeled with their name to class after three weeks and providing a written description about what they did and, importantly, how they know what they did made a difference. Challenge the students, "Convince me that you know what caused your plants to grow as tall as they did—your plants may be taller than others, but how do you know that what you did to your plants made them grow taller? (Could they just be "tall" plants)?" Have students bring to class on the day plant height is measured written answers to three questions: (1) How many pots of plants did you grow?, (2) What did you do to the pots and the plants? (What were your methods?), (3) How do you know that what you did actually made a difference? (How do you know what caused your plants to grow as they did?). Many students will take two different pots but will treat the plants in both pots the same way. Thus, they have no control or means of comparing the results of their "treated/experimental" plants.

In keeping with the spirit of a contest, set rules for judging and awarding points. For example: Students will all measure their plants using a ruler. The heights of all plants (i.e., up to five) in the same pot will be added together as the *total height* of the plants. Students with plants in the highest one-third of all total height values for the class are **eligible** for the **maximum** number of points. Students bringing in live plants not in the top third will receive **half** of the maximum number of points, while students not bringing in any live plants will receive **no** points. The contest could be held within a single class, but it has also been effectively used as a competition between 8th and 9th grade science classes within a school participating in PlantingScience (www.PlantingScience.org). Bottom line: if students didn't choose two pots and use one of them as a control, then it will be difficult for them to argue that what they did to grow large plants actually is the cause of their growth.

Another way to set up this contest, if you have time, is to run two series of contests. The first time students receive just one pot and five seeds. The purpose of this first round is to provide an opportunity for students to develop some idea of what factors seem to have an effect on growth. Then in the second round, use the "How do you know" guiding questions for students and offer a repeat with two pots.

What Makes a Good Experiment?

In a nutshell: This activity focuses on several aspects of designing experiments that are particularly challenging for inexperienced investigators: isolating just one variable to test at a time, and asking a question that is biologically meaningful.

A newspaper article describes an award-winning experiment on the effect of music on plant growth and development that was conducted by a young student. The student's project was approved by her teacher, awarded a prize at the local science fair, and reported on by a local newsman. The description of the project is as follows: The project design was to plant three pea seeds in identical soil and keep them in like environments with identical care. They differed only in the types of music and lighting to which the plants were exposed. The experiment lasted three weeks. Peas exposed to Christian music were placed in direct lamplight. Pea plants exposed to easy listening music were left in an open lighted area on a lab bench. Peas exposed to rock and roll music had indirect fluorescent lighting, and peas exposed to country music were placed in direct sunlight near a window. The results of the experiment were that the peas exposed to Christian music flowered first, had the largest leaves, and grew the tallest. Peas exposed to easy listening music were second in size, followed by the country and western peas. Peas exposed to rock and roll music died.

Have your students read about the project and ask them what they think about the experimental design—is it appropriate for investigating the effects of music on the growth of plants, and why or why not? If they were to change parts of the experiment, what would they do, and why? This student certainly discovered that the peas in the different conditions grew differently. What do they think was the actual cause of these differences? Most students will say that the light was what caused the differences, however, because the experiment was so

Inquiring About Plants

poorly designed—too many variables—there is no way to tell if the music or the light had an effect on the growth of the plants.

Plenty of similar student projects are available online for your students to review and discuss in addition to the science fair winner described above. Playing music to plants is a favorite of young investigators who start from the naïve conception that plants sense the world as they do. Here is another situation in which being intentional to have students consider life from a plant's perspective can help overcome common stumbling blocks. Ask students, "How do plants sense their world?" Ideas offered by the class could be followed up with research or readings. For example, *What a Plant Knows: A Field Guide to the Senses* written by David Chamovitz for the public would be accessible to students across a wide range of ages.

Resource:
http://www.miniscience.com/projects/plantmusic/index.html

What's the Most Popular Car in Your Town?

In a nutshell: An advantage of this activity is that it permits authentic hypothesis testing without requiring background knowledge in biology. It is thus suitable for a first-week activity to introduce making predictions, collecting data, and calculating summary statistics. Intentionally introducing students to observational studies also serves the purpose to address common misconceptions that experiments are the only valid type of scientific study.

This exercise starts with a question, "What is the most popular make of car in (name your town)?" Student teams should brainstorm for about 10 minutes to determine what might be some possible ways to determine the most popular makes within a one-hour time limit. State that the Internet cannot be used as a source of information. The usual examples are parking lots (student vs faculty vs local store, etc.) and drive-by counting on a street (one way or both directions, time of day). Go round-robin through the class groups to list possible data collection methods. This provides an opportunity to discuss random sampling, constructing a data collection table, and standardizing sampling size.

Assign each group one of the sampling methods and set a time limit for when sampling should be completed (for instance, be back to class within 20 or 30 minutes or bring data to class the next day). Ask each group to report back their five most popular makes, with total number for each. Encourage students to discuss how results across the groups compare—how similar or different are the groups' results. Compile these data into a class summary data table.

From the class summary data choose the three most popular car makes and calculate the average number of vehicles for each, plus or minus the standard deviation. Plot a histogram of your data, including the error bars. Sample class data table and graphs are below. If error bars overlap, this suggests that apparent differences are probably not significant and the differences are due to low sample size and/or high variability. If the error bars do not overlap, this suggests that apparent differences may be significant depending on the size of the differences. There are sometimes noticeable differences between parking lots devoted to specific demographics. This provides an additional opportunity to talk about bias in data collection.

Table 1.1 Class data table for cars sampled in parking lots.

Car Make	Lot 1	Lot 2	Lot 3	Sum	Mean	Standard Deviation	Standard Error
Chevy	23	23	19	65	21.7	2.3	1.3
Ford	19	28	10	57	19	9	5.2
Toyota	20	27	7	54	18	10.1	5.9
Dodge	16	14	4	34	11.3	6.4	3.7
Honda	8	6	2	19	6.3	3.8	2.2
Nissan	2	12	5	19	6.3	5.1	3

Figure 1.9 Makes of cars sampled in parking lots (mean and standard error).

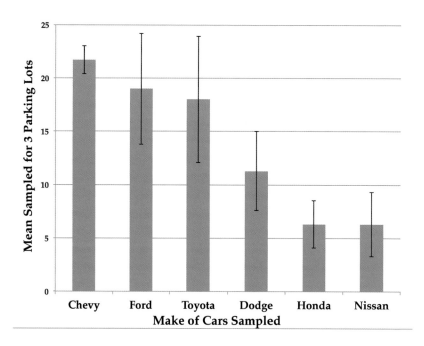

Inquiring About Plants

A Botanical Extension: Now the question becomes, "What are the most common trees along streets in this town?"

This could be done as a homework (over the weekend) project by individuals or small groups. Again, it begins with some brainstorming about what data to collect and how to collect appropriate data. Students might be asked to sign up to survey a block near their home. Some alternatives might be to collect tree data a certain distance along a street or road (one side or both), sections of streets, or blocks chosen randomly off a map of the city, or chosen specifically to target certain areas. The restriction is that only trees on public property can be sampled, between the sidewalk and the street or within 10 ft of the street.

Resource:

Quantitative Skill Guide by the College Board's Advanced Placement Biology Group

http://media.collegeboard.com/digitalServices/pdf/ap/AP_Bio_Quantitative_Skills_Guide-2ndPrinting_lkd.pdf

5) COLLECTING AND ANALYZING DATA

Analyzing Plant Diversity

In a nutshell: This activity carries over from the botanical extension above to focus on the type of data that could be collected to answer the question about which trees are common.

This is also a great opportunity to give your students practice in collecting and identifying plants. The simplest data collection method is to collect leaf samples and label them by species types (species 1, species 2, species 3, etc.), tallying how many of each species was observed. Use a marker or pen to number directly on the leaves, which can be collected in a plastic bag. When students bring their collections back to school, they can be laid out by species type to obtain class totals.

A more rigorous alternative is to have students identify the species involved as described in the previous activity. One way to help students easily identify plants is to press and dry the leaves, branches, and flowers (if any) of local plants. Take one representative sample of each species of dried plant and tape the plant to a piece of paper. Write identification labels on this sheet, noting what kind of plant it is and any other information you might want (e.g., location, ecology, when the plant flowers). Then photocopy the sheets of the local plants and place them in binders. These pictures of plants can be taken out into the field or kept in the laboratory and used as a quick way for students to match a living plant to the picture of the identified plant on the copied sheets.

After students identify the species they collected, the class data can now be used to calculate means and standard errors for each species. There are several online keys for plant identification that are useful; see especially the National Arbor Foundation resource.

Figure 1.10 A selection of student plant collections.

Data collected in this way can also be used to calculate the frequency and density of each species. Frequency and density are ecological terms used to describe plant communities. The frequency of a species is the number of individual samples containing that species divided by the total number of samples taken. For instance, if you had 10 groups of students and three of the groups had cottonwood trees in their sample area, the frequency of cottonwood is 3/10 = 0.30. The relative frequency is the frequency of one species divided by the summation of the frequencies of all the species. For instance if cottonwood was 0.30 and the summation of 19 additional species found by the class was 3.00, then the relative frequency of cottonwood is 0.10. If cottonwood was the only species present in town, the relative frequency would be 1.0, that is 100%.

The density of a species is the number of individuals of that species per unit area. For instance, if each of the 10 groups sampled a block that was 1/100 of a square mile, the total area covered would be 1/10 of a square mile. If there were a total of 15 cottonwoods in the sample, the density of cottonwoods would be 150/square mile. The relative density of a species is the density of that species divided by the total density of all species. Again, if cottonwood was the only species present in town, the relative density would be 1.0, that is 100%.

Relative frequency and relative density are two components of a species that are used to determine its importance in the community. The greater these two values, the more important the species is to the community. (There is a third component to the importance value, relative coverage, but this calculation requires additional measurements in the field).

Resources:

What Tree Is That?: Tree Identification Field Guide from Arborday.org
   www.arborday.org/whatTree

Leafsnap: An electronic Field Guide
   http://leafsnap.com

Several states also have keys specific for their state. One of the best is for
   Ohio: What Tree Is It?
   www.oplin.org/tree/

Sharing the Colorado River

In a nutshell: This activity is based on a real situation that continues to have a major impact on the entire southwestern region of the U.S. and adjacent Mexico. It dramatically emphasizes the importance of having multiple repetitions of data to increase reliability.

In the following, suggested teacher questions are in roman, *possible student answers are in italic type*, and background information to interpret the questions are in roman sans.

•▸ The Colorado River is the major source of freshwater for a watershed of seven southwestern states and adjacent Mexico. In the early 20th Century, California was growing rapidly and the other states realized that their future development depended upon water allocation from this river. In 1902, the Reclamation Act provided for development of irrigation projects throughout the West, including projects in California, Arizona, Utah and Colorado in the Colorado River Basin.

   A commission was established to determine how to equitably share the river water. Assume you were a member of that committee. Talk over in your group what important pieces of information you would need in order to make a decision.

   Typical student answers include responses such as: how many people are there in the region?; where are they located?; and what do they use water for? Occasionally a student will ask how much water is there to divide? But usually this has to be drawn out with additional leading questions.

•▸ One of the first things the commission decided they needed to know was how much water typically flowed down the river? Brainstorm in your group how you would go about determining how much water flows down the river.

Typical answers include: how much does it rain and/or snow every year and putting a gauge in the river. But we want to be more specific and you'll usually have to draw that out.

Figure 1.11 Map of the Colorado River and its drainage (yellow).

One thing that came up is to use a gauge to measure the amount of water. If you had a gauge, it would measure how much water flows through it in a certain period of time. How could you get the same information if you did not have a gauge?

Toss a float (stick) in the water and record how long it takes to float a measured distance, like 100 ft.

Will it make a difference where in the river you put your meter or throw your stick? How?

Some students will know that water flows at different rates in different parts of the river. Generally, it will be faster near the middle than near either shore and faster on the outside of a curve than on the inside of a curve.

So, if it does make a difference, where should we measure?

The idea we want to draw out is that multiple measures are necessary and then an average can be determined.

Inquiring About Plants

✦⊦ So now we have determined that at any point along the river the rate of flow will vary, depending on which part of the river is measured, so we have to take multiple measurements and calculate an average rate. But what else do we still have to know in order to answer the original question—how MUCH water flows past us down the river? So far we've only determined how FAST it's flowing.

Many students will realize that to determine the volume, you need to know something about the depth and width of the channel.

✦⊦ How can you determine the depth of the water and use that to help figure out the flow?

Measuring the depth at different distances across the river will allow you to draw a cross-section of the river's depth as in this diagram. With a little geometry you can now calculate a cross-sectional area of the river. Area times the rate will let you calculate volume per unit time.

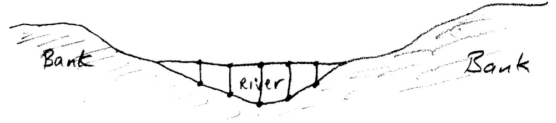

✦⊦ We now have a way to measure the volume of water flowing past a point on the river, but when should we measure it? Is once enough?

Figure 1.12 Sketch of a cross-section of the river and river banks.

Several students will probably mention that the depth probably varies with the seasons but others will suggest X-number of times for no particular reason. Try to get the class to reach a consensus of about 12 times—once a month.

Once a month is a reasonable answer that should take into account the variability of water flow throughout the year. In fact, that is what the commission ordered to have done!

✦⊦ But where should these measurements be made? The river is long and goes through several states before it crosses the border into Mexico and finally reaches the Gulf of California.

Many students will say near the mouth of the river or at the border with Mexico. What we want to lead to is several gauging stations along the length of the river.

✦⊦ The commission required that several sites along the river be monitored throughout the year so that a better record of river flow could be established throughout the basin.

The Colorado River Compact (1922), the first of a series of laws and agreements specific to the Colorado River, divided the river into an upper and lower basin. It allocated 7.5 million acre feet (maf) of water to each basin.

> That is, enough water to cover 7.5 million acres with a foot of water. The states in each basin negotiated their "fair share."

•➤ In 1944, Mexico was granted an allocation of 1.5 maf and a subsequent agreement specified allowable salinity (See allocation in the following table).

> The lower basin has priority, so in dry years the upper basin allocation must be cut back to ensure 7.5 maf flows to the lower basin.
>
> The courts have allocated 0.9 maf to Native Americans in the lower basin, and Native American rights in the upper basin are pending.

•➤ Take a few minutes to discuss with your group what you think is the biggest potential problem given the data in the table?

> Occasionally some students will see that the allocation, by law, is significantly greater than the average flow down the river. Usually smaller points are raised—Native Americans were not originally included, California is over allocation, Mexico is at allocation, etc.
>
> The single biggest problem is that, legally, 16.5 million acre feet are allocated, but the average annual flow for the past 300 years is only 13.5 million acre feet.

•➤ How could there be such a big discrepancy between the water flow calculated by the Commission, which was used to write the laws, and the actual average flow?

> *The usual answer is that mistakes must have been made in measuring or calculating the river flow in the early 1900s. Occasionally a student will suggest that maybe it was a wetter than normal year.*

> In the winter before river flow was determined, there was much higher than normal snowfall throughout the mountains of the western U.S. As a result, river flow was unusually high.

Table 1.2 The water usage data (million acre feet per year) for the Colorado River.

Region	Allocation	1990	2000	2010	2020
Upper Basin					
Colorado	3.9	2.3	2.4	2.6	2.6
New Mexico	0.9	0.5	0.5	0.7	0.7
Utah	1.7	0.9	1.0	1.0	1.2
Wyoming	1.0	0.5	0.5	0.5	0.5
Total Upper Basin	7.5	4.2	4.4	4.8	5.0
Lower Basin					
Nevada	0.3	0.2	0.3	0.3	0.3
Arizona	2.8	1.3	2.0	2.4	2.6
California	4.4	5.2	4.9	4.8	4.6
Total Lower Basin	7.5	6.7	7.2	7.5	7.5
Mexico	1.5	1.5	1.5	1.5	1.5
Grand Total	16.5	12.4	13.2	13.8	14.0

Notes on allocations:
1963 Supreme Court Ruling on Lower Basin allocation to Native Americans: 0.9
Pending ruling on Upper Basin allocation to Navajo Nation: 2–5
1954–1963 average of water flow: 11.8
Range of water flow: 4.4 to >22

●▸ Discuss for a minute what you could have done to avoid this problem if you were a member of the commission.

What we're looking for is multiple years of data to get an average flow.

6) WORKING WITH REPRESENTATIONS (GRAPHS, DATA TABLES, ETC.)

There are many opportunities for students to plan for the collection of data, and to work with and interpret the data they have collected. While most people think that these opportunities are restricted to activities related to laboratories, there are, and should be, many chances for your students to work with data sets and their interpretation throughout your class.

If you are interested in getting your students to use and understand statistical analyses, you might refer them to the new Quantitative Skills Guide produced by the College Board's Advanced Placement (AP) Biology group:

http://media.collegeboard.com/digitalServices/pdf/ap/AP_Bio_Quantitative_Skills_Guide-2ndPrinting_lkd.pdf

You can help students with their data interpretation skills by providing them with data sets and asking them to construct an appropriate graph that illustrates those data in a meaningful way—i.e., should the data be represented as a bar graph or a line graph, how should the X- and Y-axes be labeled, and what should the title of the graph be? Additionally, you can provide students with a graph and ask them to interpret it, articulating what the graph is "telling them." You can find a wide variety of graphs from scientific literature, from textbooks, from newspapers and magazines, and certainly from the Internet. Some of these graphs, and data sets, may not be accurately constructed, or may be misinterpreted, which provides a rich opportunity for discussion. As you build up your repertoire of activities, consider exposing students to graphs that illustrate data in "atypical" ways to stretch their data interpretation skills. The following section on energy allocation in roots includes an "unusual" graph.

Figure 1.13 Germinating seeds.

Seeds and Seedling Growth

Ask students to make observations about the picture of a germinating seed. What do they notice? What part(s) of the seed emerge first when the seed germinates? What is the significance of this? (What does this suggest that the seedling requires?) Where is the embryo getting the energy it needs to produce plant parts before it can start to photosynthesize on its own?

Inquiring About Plants

A scientist wanted to see how extensive the root system of a plant was, so he planted a single rye seed in a box with the dimensions of 1' X 1' X 2'. The seed germinated and grew for 4 months before the rye plant was carefully removed from the box and the soil washed from the root system. The scientist then estimated the number and length of the roots and root hairs of this one plant. Here are the resulting data from that study.

Have students review these data before you ask them questions—but you need to get them to notice the incredible numbers of roots (in the MILLIONS) and root hairs (in the BILLIONS) of this single plant and their incredible lengths (MILES long). Remind them how big (tiny) the container was in which the 4-month old plant grew!

Table 1.3 Tables 1 and 2 from Dittmer's original article.

A. ROOTS.

Root categories	Number of roots per unit[a]	Total number of roots by categories	Average length (inches)	Total root length by categories (feet)	Average diameters (microns)	Total root surface by categories (sq. feet)
Main	1.0	143	18.0	214.50	700	1.53
Secondary	249.0	3.5,607	6.0	17.803.50	250	45.06
Tertiary	16,060.5	2,296,651	3.0	574,162.75	130	758.60
Quaternary	80,302.6	11.484,271	1.5	1,452,075.60	120	1,748.90
Totals	97,613.1	13,815,672		2,044,256.50		2,554.09

[a]A unit includes one main root with all its secondary, tertiary, and quaternary roots. There were 143 such units on this plant.

B. ROOT HAIRS.

Root categories	Number of hairs per sample[a]	Total number of root hairs by categories	Average length of root hairs (microns)	Total root hair length by categories (miles)	Average diameters (microns)	Total root hair surface by categories (sq. feet)
Main	53.25	3,481,46.1	1,000	2.16	15	1.73
Secondary	45.00	244,196,860	860	130.20	12	87.65
Tertiary	33.00	5,775,159,861	800	2.864.40	12	1,873.72
Quaternary	19.00	8,312,730,104	700	3,607.10	12	2,359.94
Totals		14,335,568,288		6.603.86		4,321.31

[a] The number of hairs on a root length of 1 mm.

Summary from the original paper: A quantitative study was made of the number, length, and total exposed surface area of the roots and root hairs of one rye plant (*Secale cereale* L.). Counts were made by categories (main, secondary, tertiary, and quaternary) to determine the total number of these roots by ranks. The 13,815,672 roots had a surface area of 2,554.09 square feet. Diameters of roots were characteristic of the categories to which they belonged. Living root hairs on this plant numbered 14,335,568,288 and had a total surface area of 4,321.31 square feet. They covered all roots of each category, but they occurred in greatest number on roots of the quaternary division. The root hair surface combined with that of the roots gave a total of 6,875.4 square feet, which was the area of possible soil contact for this plant. The total external surface of the 80 shoots with their 480 leaves was 51.38 square feet; the surface area of subterranean parts was therefore 130 times that exposed by the top.

You might ask students to figure out a way they can illustrate how much surface area this one root system possessed (e.g., compared to the area of a basketball court). Ask your students if there is a better way to illustrate these data, and different ways to interpret them—are there correlations between any of the data in different categories?

Energy Allocation

Figure 1.14 Energy allocation of an annual plant, *Senecio vulgaris*, measured as percent allocation of dry weight to different structures throughout its life cycle. (From Harper, 1970)

Now, here is an interesting graph of the energy allocation of an annual plant, *Senecio vulgaris*.

While it is unrelated to the rye plant, both *S. vulgaris* and *S. cereale* are annual plants, and so they share the same life history. They both start from seed, germinate, grow and reproduce within a period less than one year. And. . . . they only get one chance to reproduce (as compared to a perennial, which can reproduce each year after it reaches maturity). So, if a plant can only reproduce once, how much energy do students predict the annual plant will allocate to its roots versus its reproductive parts, such as flowers, receptacles, and seeds?

It will take students time to determine how the graph above is set up—notice that the X-axis is time, directly related to the life of this annual from its seedling stage to its maturity, when it flowers and sets seed.

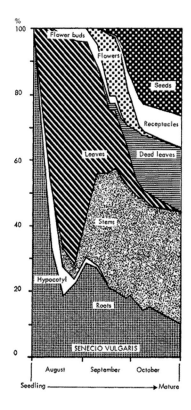

Inquiring About Plants

The Y-axis is dry weight, which is a measure of how much energy the plant has allocated to particular parts, at different times in the life of the annual plant. Thus, in its seedling stage, all of the energy is going into the production of roots (which agrees with the observations students made of the germinating seed above). And, nearing the end of its life, the annual allocates much more energy into its reproductive parts (receptacles and seeds) than into the roots, which makes sense if one considers that this plant only has one chance to pass its genes on to the next generation. So, one challenge students will have is interpreting this kind of graph. Get them to predict what the graph might look like for both a biennial plant (such as a carrot that stores a lot of energy in its root system during the first year of its life) versus that for a perennial plant.

In addition, challenge students to reconcile these two ideas—the huge root system produced by annual plants (such as the rye plant) with the small amount of energy that annuals typically allocate to their root systems (e.g., *S. vulgaris*), understanding that we are comparing different species. The question is: "How can annual root systems become so large with so little energy, comparatively, being allocated to the roots?" The answer can be found in the tables above—the roots of these plants are very thin and the root hairs much, much thinner (a few microns wide). Thus, the annual doesn't have to use much energy to produce very long, very thin roots or root hairs—producing a huge root system with as little energy as possible.

You might consider adapting/adopting some of the plant biology investigations that can be found in the College Board's Advanced Placement (AP) Biology Laboratory Manual. All of these investigations are designed to start off students with some directions, but end up with students designing their own experiments based on their own questions. One investigation is about transpiration, and it allows students to collect and work with data that they collect and with datasets that have been collected by others.

Resource:

The College Board's Advance Placement Biology Laboratory Manual
http://media.collegeboard.com/digitalServices/pdf/ap/bio-manual/CB_Bio_Lab_11_WEB_1_24_12.pdf

Stomata

In the first part of the transpiration investigation, students collect data about the number of stomata on one leaf (comparing the top and the bottom surfaces of a leaf), and then they compare the number of stomata of leaves from different plant species or different places on a single plant—such as sun and shade leaves on a tree. Students determine the average number of stomata per square millimeter in a particular kind of plant.

Several questions should arise about stomata, including the function of stomata and their control, but mainly..."How are the density of stomata and the size of a leaf related to how much water a plant loses through the stomata?" Also:

- If a plant loses too much water, it will wilt and die, so why wouldn't a plant have any stomata, let alone millions?
- Do all plants have stomata? Do all plant parts have stomata?
- What might happen to the rate of transpiration if the number of leaves or the size of leaves is reduced?
- Do all plants transpire at the same rate?
- Is there a relationship between the number of stomata and the environment in which the plant evolved?

ESTIMATING STOMATA NUMBER

To estimate the number of stomata on a leaf, students can make a wet mount using nail polish that is first painted on the leaf epidermis.

1. Students choose a leaf (almost any kind of leaf can be used, however, one that is smooth and not hairy is a good place to start).
2. They should paint a solid patch of clear nail polish on the leaf surface (either top or bottom side of the leaf). They should make a patch of at least one square centimeter and allow the nail polish to dry completely.
3. Using forceps carefully pull the nail polish patch away from the leaf, place it on a microscope slide, and then prepare a wet mount of the peel using a drop of water and a cover slip.
4. Students should examine the leaf impression under a light microscope at 400X (or highest magnification), and draw and label what they observe in areas where there are numerous stomata.
5. They should count all the stomata in one microscopic field of view, and record the number. Repeat counts for at least three other areas of the same patch, and calculate the average number of stomata per microscopic field.
6. Using this average, they should be able to calculate the number of stomata per 1 mm^2. They can estimate the area of the field of view by placing a transparent plastic ruler along its diameter, measuring the field's diameter, and then calculating area by using πr^2. (Most low-power fields have a diameter between 1.5–2.0 mm.)
7. Ask students to create a table to illustrate the data they have collected from different leaves or parts of leaves.
8. Now, ask them to compare their data with the data in Table 1.4.

Table. 1.4 Average number of stomata/mm² of leaf surface.

Plant	In Upper Epidermis	In Lower Epidermis
Anacharis	0	0
Coleus	0	141
Black Walnut	0	460
Nasturtium	0	130
Oats	25	23
Corn	70	88
Tomato	12	130
Sunflower	85	156
Water Lily	460	0

To which of the above plants did the students' plant most closely resemble? Do those two plants live in similar environments, or how might students explain their similarity? How do students explain the number of stomata in *Anacharis* (Hint: it's an aquatic plant that lives entirely underwater), or water lily (Hint: the leaf blade sits on the surface of the water, so if it had stomata on its bottom side, they would be completely blocked), or corn (Hint: think about how corn leaves grow—do they have a clearly defined top and bottom?).

Here is another type of representation, a histogram on changes in the density of stomata on plants that have been exposed to higher concentrations of carbon dioxide. What is the relationship between the concentration of carbon dioxide in the environment and the density of stomata? How does that relate to the function of a stoma? For an interpretation of this graph and additional graphs and tables, see the original article by Woodward and Kelly: (Responses of the 100 plant species surveyed followed close to a bell curve and 75% showed reductions in stomatal density with CO_2 enrichment).

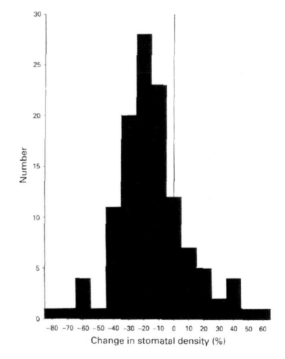

Figure 1.15 Stomatal density responses to CO2 enrichment, measured as percentage changes relative to the lowest CO2 concentration studied. (From Woodward and Kelly, 1995)

Evolution and Measuring Variation

In a nutshell: This activity uses careful observations and graphing to demonstrate variability among individuals of a species. The student task to measure
variation using sunflower seeds is connected here to foundational ideas
about how evolution works. For an elaboration, descriptive statistics and/or a
simple t-test can be used in comparing the results of different groups.

In the following, suggested teacher questions are in roman, *possible student
answers are in italic type*, and background information to interpret the questions
are in roman sans.

> *Rationale:*
> Your students may have heard or read about the French botanist, Lamarck,
> who proposed a theory of evolution before Darwin. Lamarck's mechanism
> to describe how evolution occurs was not supported by the evidence. How-
> ever, he did have one idea that was picked up by Darwin and is essential to
> understanding the Theory of Natural Selection.
>
> Part of Lamark's theory of evolution was that there must be variation among
> individuals of a species. His reasoning was this. If species were created, the
> creator would have made them perfectly suited to their environment and
> there should be no variation between individuals: *"Nature (or its author) in
> creating the animals, has . . . given to each species an unchanging orga-
> nization, as also a form determinate and invariable in its different parts."*
> On the other hand, if evolution occurred, individuals should vary from one
> another *"Nature . . . has gradually complicated their [species] structure . . .
> and modifications in its organs."* Variation between individuals is also a key
> component of Darwin and Wallace's Theory of Natural Selection.

The French botanist Lamarck proposed that if species were created "in a form
determinate and invariable" there should be no variation between individuals.
On the other hand, if species evolved, there should be variation between individuals. Which of these two hypotheses is supported by observations of sunflower
fruits?

Pass out sunflower seeds for student groups to examine. (Note: what the general public calls sunflower seeds (in the shell) are technically fruits called achenes. Also, pecans or other fruits can be used).

•▸ Within your group, make a list of at least five measurable characteristics of these fruits that you could measure.

Figure 1.16 A selection of pecans and sunflower seeds in the shell.

> *After five minutes, have the groups contribute to a class list of characteristics you can put on the board. Characters will usually include: length, width, thickness, mass, volume, surface area, and color pattern. Discuss with the class how each of the characteristics they mentioned could be measured.*

•▸ Let's take the characteristics you've identified and discuss how each of them can be measured.

Length: Length is a straight forward measurement and can be measured quickly. A metric ruler will work but a vernier caliper would be better.

Width and Thickness: These measurements are not as straight forward as students might think. Where should the measurement be taken? Both width and thickness will vary along the length of the fruit. Ask students where they could measure to get the most consistent and accurate readings—the widest point or the thickest point works best. Use a metric ruler or vernier caliper.

Mass: Mass is also straight forward and can be measured quickly using a balance.

Volume: The easiest way to measure volume is by water displacement. Use the smallest graduated cylinder that the fruit will fit inside. Fill it to a mark with water. Then place the fruit in the water. If it floats, use a dissecting needle to submerge it just below the water. Read the new water level.

Surface Area: Surface area can be measured in several ways. The easiest is to trace the outline of the fruit on a fine-ruled metric grid paper. Count the number of squares completely within the outline plus ½ the number of squares partially included. Multiply this by the area of a single square. Alternatively, the outline can be traced on plain paper, then cut out and weighed. Then, cut out a square of known size from the same paper and weigh it. The weight of the outlined paper divided by the weight of the known sized paper will tell you the fraction of the known area.

If you have digital cameras and computers, you can use Image J.

Color Pattern: Make a transparency of the fine-ruled metric grid paper used for surface area. Lay this over the seed and count the number of squares completely covering and ½ the number of squares partially covering each of the colors patterned on the fruit coat.

•▸ Before we get started, we'll have to decide how many measurements is enough to determine if there is variability in each of these characteristics. Do you think one is enough? If not, how many?

Ten is probably a good number, just to keep the math easy, but in an advanced class you will want to teach them how to determine the minimum sample size for an investigation. One way to do this is a modified "species area curve" from ecology. For instance, start by making two measurements of the parameter of interest, e.g., length. Calculate the mean and standard error of these two. Now add two more measurements and calculate the mean and standard error for all four. Add two more and repeat the process for all six. Repeat this process as often as necessary. When the sample size is large enough, the mean will stabilize and the standard error will minimize. This will be easiest to see if the data are graphed— the mean will level off, and the standard error bars will be small and constant.

Assign each group one parameter to examine and measure. They should record their data in a table, then construct a bar graph of their data showing how many individuals fall into each size class from the range of observed values. Sample data are shown below.

Figure 1.17 Sample class data showing distribution of length and mass size classes.

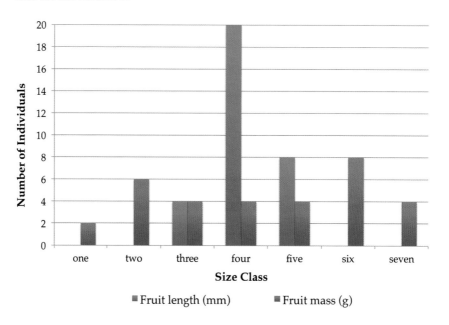

Inquiring About Plants

Have each group put their graphed data on the board, then interpret their results relative to the hypothesis that there is no variation between individuals. Do the data support this hypothesis?

Some points to emphasize:

- Different characters will vary by different amounts. For instance, in the example above, fruit mass varies more than fruit length. How would two scientists' conclusions about variation compare if one measured only width and the other only mass? Would one be more biased than the other?
- You have to accept your data (unless there is an obvious reason why it should be questioned, such as experimental error—and even then the experiment should be repeated, not the interpretation reversed). If the data require you to reject the hypothesis, you must reject it, no matter how much this goes against your "common sense." Of course you will want to try an alternative way of testing your hypothesis to be sure the experiment you tried wasn't biased.

COMPARING CHARACTERS

As suggested by the sample data above, the results from measuring different characteristics will probably be different. It may be interesting to ask if these differences are great enough to say that the results are significantly different. It may also be interesting to compare the results of two different classes. For instance, we could compare fruit lengths obtained by one class with fruit lengths measured by another class.

•» Both sets of data on the graph above support the idea that there is variability among individuals, but they seem to show two different patterns. Do you think these patterns are different enough to be considered significant?

In science, using the term "significant" implies that a statistical test was performed. An easy way to demonstrate this is by comparing the class's two most strikingly different sets of data by using a t-test. Because each group had the same sample size you will have paired samples. You will need the raw data for size classes for each sample. For instance, the data in Figure 1.17 for fruit length are: 3, 3, 3, 3, 4, 4, 4, 4, 4, 4, 4, 4, 4, 4, 4, 4, 4, 4, 4, 4, 4, 4, 5, 5, 5, 5, 5, 5, 5, 5 and the comparable date for fruit mass is: 1, 1, 2, 2, 2, 2, 2, 2, 3, 3, 3, 3, 4, 4, 4, 4, 5, 5, 5, 5, 6, 6, 6, 6, 6, 6, 6, 7, 7, 7, 7. Excel can be used to calculate the t-statistic and critical value (t-test for the mean of paired-samples). Alternatively, "t" can be calculated by hand and the resulting value compared to a table of critical values.

Resource:
ImageJ (Image Processing and Analysis in Java): http://rsbweb.nih.gov/ij/v

8) WORKING WITH SCIENTIFIC EVIDENCE AND ALTERNATIVE EVIDENCE-BASED EXPLANATIONS

The Darwins and Phototropism

In a nutshell: This activity builds on observation and hypothesis testing skills. More important, however, is that it illustrates the importance of always questioning the conclusions you draw from an experiment—is there an alternative explanation for my results? This will help to dispel the idea that once you complete an experiment and analyze your results, you are finished.

In the following, suggested teacher questions are in roman, *possible student answers are in italic type*, and background information to interpret the questions are in roman sans.

Observe the QuickTime video on corn phototropism at the following site: http://plantsinmotion.bio.indiana.edu/plantmotion/movements/tropism/tropisms.html

•▸ Assume one of your classmates is unable to view it. How would you describe what is shown in the video?

> *Initially it looks like little leafless stems standing straight up out of the ground in a circle surrounding a light bulb. When the light bulb is turned on, all of the "stems" begin to bend toward the light. When that bulb is turned off and the room lights are turned back on, the "stems" all straighten back up.*

•▸ Bending toward light is known as phototropism. This response of plants was documented by the ancient Romans. However, it wasn't until the late 1800s that this phenomenon began to be investigated scientifically.

The investigators were Charles Darwin and his son Francis. Francis was a physician trained in physiology. He was so intrigued by the experiments with his father, published in *The Power of Movement in Plants*, that he spent the rest of his career as a plant physiologist.

The Darwins used grass seedlings for their studies because they grow so rapidly and respond even before the first true leaves emerge from their covering, the coleoptile. Below is an image of a wheat plant, less than a week old. The plants were exposed to light from one side. Which side do you think this was, and what is the explanation for your choice?

The light was coming from the right side.

⇢ The question for the Darwins was "what part of the plant senses light?" This is the necessary first step in bending towards the light. Based on your observations above, brainstorm in your group about what are some possibilities about plant parts that might be light sensitive.

> *The bend is actually toward the base, so this might be where light is sensed.*
> *The tip is the furthest bent so this might be where light is sensed.*
> *Light could be sensed anywhere along the length.*

⇢ The Darwins actually considered and tested each of these hypotheses, but let's limit ourselves to the tip. Time for more brainstorming: what is the simplest experiment you could do in the lab to test if the tip is the part of the plant sensitive to light?

> *Cut the tip off of some plants, but leave others intact as a control.*

Figure 1.18 Germinating wheat seedlings with green coleoptile (cylindrical first leaf) emerging from two of the seeds. The white root emerges first, visible on the right-hand seedling.

Fig. 1.19 Comparison of wheat seedlings. A. Intact seedlings. B. Seedling with tip of coleoptile removed.

↠ These photos illustrate the results of one of the Darwins' first experiments to determine if the tip is the part of the seedling that senses light. The plants with the tips removed did not bend, but the control plants did. Does this support the hypothesis that the tip senses light?

Yes, it supports the hypothesis.

↠ While this provided support for their hypothesis, the Darwins realized that there was another possible explanation that would give the same results. Take a minute in your group to come up with a possible alternative explanation for the decapitated seedling not bending.

It may take some prompting for some groups to come up with the idea that the physical injury may have inhibited the response. One way to elicit this is to ask a student volunteer to stand and turn to face light coming in a window. Ask the volunteer how s/he sensed the light. How would physical injury to the eyes affect how a person could sense and respond to light?

↠ The Darwins realized that the physical trauma of decapitation could be a possible explanation why the plant on the right did not respond. Take a minute in your group to discuss how you could use some commonly available materials in the lab (or kitchen) to modify your experiment to test if the tip is sensing light without decapitating the plant.

Figure 1.20
Comparison of wheat seedlings. A. Uncovered seedlings. B. Seedling with aluminum foil "cap."

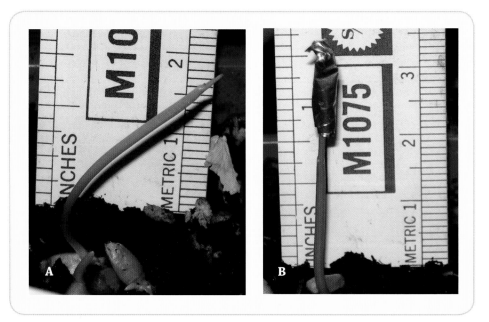

Inquiring About Plants

The two most common answers are to paint the tip or to cover it. Painting could work, but with what? Lamp black, India ink, and black enamel were all available and used by Darwin, but there's always a toxicity problem to consider. An easier approach would be to cover the tip with something light proof. Aluminum foil would work in the lab. Darwin made tin foil caps (aluminum foil was not yet invented).

This image illustrates the results when Darwin covered the tips of some seedlings with tin foil caps. A small piece of foil was wrapped around each treatment tip and then squeezed tight enough so no light could leak in through the bottom. Do these results support the hypothesis that the tip senses light?

Yes, only the plant with its tip exposed to light bent.

While this provided support for their hypothesis, the Darwins again realized that there was another possible explanation that would give the same results. Take a minute in your group to come up with a possible alternative explanation for why the foil-wrapped seedling did not bend.

Students often suggest that maybe the foil was too heavy. Some may suggest that because it was crimped tight around the tip, the physical constraint was keeping it from bending. Again, you may have to do some prompting. For instance, has anyone had a broken bone that was temporarily splinted with a pneumatic splint? How does a pneumatic splint keep broken bones from moving?

The Darwins were concerned that physical constraint could have prevented the tip from bending, regardless if light was being sensed. Take a minute in your group to discuss how you could use some other commonly available materials in the lab to physically constrict the tip but let light through—or not.

Transparent "Scotch" tape is usually mentioned. This might work, but it certainly wasn't available to Darwin and his son. They used glass tubing. Students may be familiar with capillary tubing. They will probably not know that you can heat glass tubing until it is glowing red, then stretch it to make the tubing thinner in diameter. This is how we make micropipettes for cell manipulations, and it is what the Darwins did to cover the tips of plants.

This is an illustration of how the Darwins eliminated the physical constraint alternative. All of the seedlings were covered with a snug-fitting glass tube. Some were painted (here with India ink) and others were left clear. The seedlings with clear

Figure 1.21 Germinating wheat seedlings covered with glass tubing "caps." Cap darkened with black India ink on left, clear glass on right, and uncovered seedling on far right.

tubing responded while the blackened ones did not. Do these results support the hypothesis that the tip senses light?

Yes they do.

→ The Darwins took this experiment even further and scratched off a thin line of paint on only one side of the blackened glass—but on different sides of different plants. In each case the plant bent toward the side where the scratch allowed light through.

In all they performed more than a dozen different experiments on phototropism using five different species of plants. Their conclusion was: "*We must therefore conclude that when seedlings are freely exposed to a lateral light some influence is transmitted from the upper to the lower part, causing the latter to bend.*"

This was the first proposal to suggest the existence of what we now call hormones that are transmitted from one part of an organism to have an effect on another part!

The Ecology Game

In a nutshell: This activity, based on a real event, emphasizes hypothesis testing and data analysis in a game-like setting. Twice students will think they have "the answer," only to discover that they do not hold up to further experiments. The take-home message is that sometimes conditions that have little individual impact can have major combined effects—synergistic effects. This game works from middle school students through graduate students in teaching assistant training. It can take between 30 minutes (a middle school class!) and 3 hours to arrive at the solution. If a class bogs down, you can provide some hints or leading questions.

In the following, examples of instructor lead-in statements or questions are in roman, *typical student responses are in italic type*, and background information is in roman sans.

The Ecology game is a problem-solving challenge based on a real event with real data. Divide the class into teams and present them with the information below. After a few minutes to brainstorm, choose a first team to propose one of their questions or hypotheses that could provide data to help them figure out what happened. They must also suggest the kind of data collection or experiment that could provide the information they want. You then retrieve an appropriate set of data from the database and show it to the entire class. It is now published so every group can use this information to refine their questions. Now invite a second team to suggest their question or hypothesis and repeat the procedure for each team. It is unlikely that any team will have a suitable explanation after the first round, so continue with additional rounds until one team reaches the answer.

To add a bit of realism about making decisions on how to use available resources to address a problem, give each team a research budget (e.g., $2,000). A dollar value is assigned to each data set in the database. Data from public records or information is free, but information that must be generated has a cost based on the amount of work that would be required to generate it. Teams must decide if they are willing to spend their money to see data they are interested in obtaining. If a team runs out of money before the answer is reached, they can only be observers.

⇥ Angle Bay (see map below) is a small resort town on the southwest coast of England. Its primary attractions are the sandy beaches with interspersed rocky boulders.

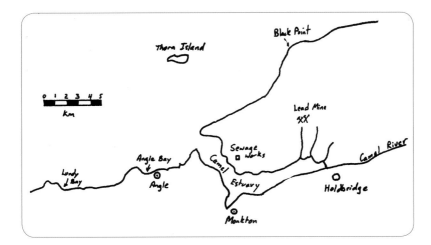

Figure 1.22 Map showing location of Angle Bay. (After Tribe and Peacock, 1976)

Figure 1.23 Angle Bay at low tide with the intertidal zone exposed.

•• Visits to Angle Bay in July 1960 and July 1967 revealed the following very obvious changes in the abundance of several conspicuous organisms:

Table. 1.5 The abundance of organisms in Angle Bay. (After Tribe and Peacock, 1976)

Observations	1960	1967
% cover *Enteromorpha* sp.	<5%	80%
% cover *Ulva* sp.	<5%	40%
Number of limpets (*Patella* sp.) / m²	180	12
Number of barnacles (*Balanus* sp.) / m²	1600	1000

•• Your assignment is to investigate this change and to uncover as much information as possible about its cause(s). The following background information should prove helpful in thinking about the problem.

Biological Background on Algae

A diverse group of organisms, algae are key primary producers in aquatic and marine environments. Their evolutionary relationships are not fully understood yet. Major groups of algae are divided on the basis of pigments present, the kind of storage products produced, features of their body type and cell wall, and other differences. The large, macroscopic algae (commonly called seaweeds) are classified as Brown, Green or Red algae.

Algae are zoned within the subtidal-intertidal area with different species characteristically found at different tidal levels. This zonation is related to the physiological tolerances of each species as well as biological interactions between algal species and between algae and marine invertebrates (competition for space, grazing).

There are two types of life cycles common among algae. Many species are annuals: they live less than one year. Many annuals die off in the autumn, and new individuals develop from spores that germinate in the spring. Growth is fast, but restricted to one season, generally spring or summer.

Other seaweeds are perennials: they live more than one year. Growth is generally slower, but continuous throughout the year. Some perennials may partially die back in autumn and winter, but basal portions regenerate in the spring.

Two of the more common green algae are:
Enteromorpha – This is a fast-growing annual, very common on intertidal rocks at all tidal levels; it frequently grows on other plants as well. It is shaped like soft, papery tubes, 2-6" long and 1/8-1/4" in diameter. Reproduction (and settlement) can take place at any time of the year.
Ulva – Sea lettuce is very similar to *Enteromorpha*, but it is a flat sheet instead of a tube.

Figure 1.24 Common algae. A. *Entermorpha*. B. *Ulva*.

Figure 1.25 Common invertebrates. A. Limpets. B. Barnacles.

Biological Background on Intertidal Animals

Most intertidal animals are invertebrates (animals without backbones). There are many phyla that may be encountered on intertidal rocks.

A few of the more common animals are:

Barnacles (Arthropod crustaceans)—These are stationary once they settle on a rock surface and rarely live more than 1–2 years. They are ½" in diameter and covered by calcareous plates. These can be opened when the animal is underwater and feathery appendages filter plankton (primarily diatoms) from the water. Barnacles reproduce in early spring and produce planktonic larvae. These float in the water column for a while, but soon select a spot on a rock surface and attach. Their main predators are drills and starfish.

Two species are common in Angle Bay: *Chthamalus* sp., which occupies the high intertidal zone, and *Balanus* sp., which occupies the mid zone.

Limpets (Molluscan gastropods)—These are intertidal snails having a flat, shield-like shell that rarely live more than 1–2 years, although a few large individuals may be 10–15 years old. They are capable of moving up to a meter per day, but rarely do so and generally return to the same place on the rock ("home"). Their shape and strong muscular foot, which enables them to cling tightly to the rocks, allows limpets to withstand violent wave action. Limpets are herbivorous, usually scraping the rock surface for diatoms and newly germinated sporelings (very small plants) and only rarely eating macroscopic (large) algae. Limpets spawn (shed gametes) primarily in winter. The larvae are planktonic for 1–2 weeks and then metamorphose into a crawling stage that settles on the rock. Their main predators are starfish and birds, especially oystercatchers.

There are several species of limpets (genus *Patella*) that from high to low intertidal zones.

⟶ Now you are ready to begin. What question or hypothesis does team 1 want to make?

Pollution from the sewerage works and lead mines are often among the first things raised.

Other factors are increase in human population, catastrophic storms, and increases in fishing. Frequently teams want data on other species of plants or animals, both from the intertidal zone and planktonic species from offshore. They also request similar data from earlier or later years and corresponding data from Lundy Bay, Black Point, or Thorn Island. Eventually someone will ask if there was an oil spill. The table and figure below are two sets of data related to that question.

Table 1.6 Oil quantities on the beach at Angle Bay from 1960–1970. (After Tribe and Peacock, 1976)

Year	Amount of oil washed ashore (in 100s of pounds)
1960	55
1961	1.5
1962	8.5
1963	8.5
1964	7.0
1965	4.0
1966	3.7
1967	550
1968	2.8
1969	4.9
1970	6.0

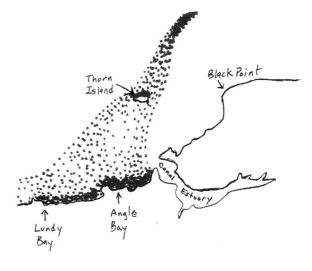

Figure 1.26 Distribution of oil on March 31, 1967. The oil is part of the cargo of 100,000 tons spilled from a tanker wreck, a week earlier, 20 miles out to sea. The area from Lundy Bay to the Camel Estuary were sprayed with 200,000 gallons of the detergent BP 1002 during the first two weeks of April to break up the oil. (After Tribe and Peacock, 1976)

⊷ So, oil is associated with the observations: it is present at Angle Bay at the right time. But association by itself does not define cause. What do you need to know?

One hypothesis is that the oil is harmful to the invertebrate species and as a result the algae will grow. To test this you need data on toxicity of oil on invertebrates.

⊷ There are data on toxicity of oil on invertebrates. This data set costs $600.

Table 1.7 Oil toxicity to invertebrates. (After Tribe and Peacock, 1976)

Invertebrates tested	Exposed to crude oil		Control	
	% mortality	Mean recovery time* (hr)	% mortality	Mean recovery time* (hr)
Monodonta sp.	0	32	0	0
Gibbula spp.	5	6	0	0
Littorina spp.	44	20	0	0
Patella spp.	10	4	0	0
Mytilus sp.	0	?	0	--
Balanus sp.	0	10	0	0
Chthamulus sp.	0	25	0	0

*Time necessary for survivors to regain normal behavior and activity. In the lab, 50 animals of each species were exposed to weathered Kuwait crude oil for 6 hours; control animals were kept in seawater. Experiments were run at 14° C.

Incorporating Science Practices in the Classroom

• Do the data support your hypothesis that the oil killed invertebrates of many or most species?

> *The oil was lethal to some of the limpets in this experiment after 6 hours, but not enough to be responsible for the observed decrease.*

> But you don't know how long they were exposed in nature and to what concentrations of oil.

• Similar data are also available on the effect of oil on algae. This data set costs $50.

Table 1.8 Oil toxicity to algae. (After Tribe and Peacock, 1976)

Algae tested	Contraction of cell contents	Loss of chlorophyll
Ascophyllum (brown)	–	–
Laminaria (brown)	+	–
Cladophora (green)	–	–
Enteromorpha (green)	–	–
Prophyra (red)	–	–

*In the lab, freshly collected algae were coated with weathered Kuwait crude oil and then immersed in seawater for 6 hours. After 6 hours, the appearance of actively growing cells was noted. ++ indicates a marked effect. + indicates some effect; – indicates no visible effect.

• Oil does not appear to have a direct effect on the observed changes in algal and invertebrate numbers associated with the oil spill. What might be another related factor? What do we usually do when there is an oil spill?

> *How was the oil spill cleaned up?*

• Here is some free information on cleaning up oil spills, including the method employed at Angle Bay.

> *We think the detergent used to clean up the spill was toxic to invertebrates.*

• Data on toxicity of the detergent used for cleanup of the Angle Bay oil spill is available for $800 (Table 1.9).

Several methods used to clean up oil spills:

1. While the oil is still at sea it may be
   a. Sprinkled with hay or chalk to make it sink to the bottom.
   b. Set on fire and burned.

2. Once the oil drifts ashore, it may be removed by
   a. Manual labor. Scraped up and gathered in bags and shipped away.
   b. Spraying with detergents. This causes it to emulsify and be washed away.
   c. Burning. This is almost impossible once the oil has weathered to any extent.
   d. Consumption by organisms. Limpets are highly effective. Attempts to breed an oil-consuming bacterial strain are still experimental.

Figure 1.27 Methods of clean up for oil spills at sea.

Table 1.9 Detergent toxicity to intertidal animals. (After Tribe and Peacock, 1976)

Invertebrate common name	Concentration of detergent BP1002 causing 50% mortality in 24 hours of exposure (ppm)	Notes
Sea anemone	25	
Crabs	36	
Shrimp	2	
Barnacles – *Balanus* sp.	130	5 ppm lethal to larvae
Worms	25	10 ppm lethal to larvae
Starfish	30	
Sea urchins	36	1 ppm causes abnormal development
Snails – *Patella* spp.	5	All snails: 1 ppm causes inactivity 3 ppm lethal to larvae
Littorina spp.	225	
Monodonta spp.	100	
Gibbula spp.	125	
Mussel – *Mytilus* spp.	70	

•→ Do these data support your hypothesis or not? What additional information do you need to interpret the data?

The detergent had some effect on the observed species, but less than on some others. We don't know how long they were exposed in nature and to what concentrations of detergent. We also don't know if larval stages were present in the plankton.

Table 1.10 Oil versus detergent: field toxicity to invertebrates and algae.
(After Tribe and Peacock, 1976)

Species	SampleTaken*	Control (No. / m²)	Oiled only (No. / m²)	Oil + Detergent (No. / m²)
Invertebrates				
Balanus sp.	Before	8300	8100	7200
	After	8000	8000	2400
Chthamalus sp.	Before	42000	40000	40000
	After	39000	37000	35000
Gibbula spp	Before	5	3	5
	After	3	2	2
Littorina spp.	Before	64	64	66
	After	65	37	7
Patella spp.	Before	125	137	130
	After	125	130	30
Algae				
Porphyra sp.	Before	4	5	4
	After	5	5	5
Enteromorpha spp.	Before	1	0	1
	After	6	10	70
Ulva sp.	Before	0	2	1
	After	3	5	25
Fucus spp.	Before	19	25	29
	After	20	30	30

*Quadrats (measured areas) were established in the mid-intertidal zone near Black Point in March 1968. Five quadrats were designated as controls; nothing was done to these. Five others were treated with Kuwait crude oil. Five others were treated with Kuwait crude oil and then cleaned 30 minutes later with detergent BP 1002. The number of living invertebrates and algal cover were measured before each treatment and again four months later. Numbers shown are means of the five quadrats.

Inquiring About Plants

••▸ It doesn't look like either the oil or the detergent used to clean up the oil can explain the results on their own. What is another possible explanation?

> Occasionally students may suggest that the combination of oil and cleanup chemicals might be involved, but usually it will take some coaching to get this request.

••▸ Data on the effect of a combination of oil plus detergent on both algae and invertebrates are available for $100 (Table 1.10).

••▸ This is an example of a synergistic effect. Neither the oil nor the detergent alone was responsible for the invertebrate decreases nor the algal increases at Angle Bay following the oil spill. But the combination of the oil and detergent was toxic to many invertebrate species and allowed rapid expansion of the algal populations.

> From additional data on invertebrate and algal populations after the 1967 Torrey Canyon oil spill, students will know that within a short time the populations returned to a "normal" state. At the time, the Torrey Canyon oil spill, 25–36 million gallons, was the world's largest. Today, it is not in the "top 10." The Gulf of Mexico Oil Spill in 2010 released about 200 million gallons (and is number 2).

Attribution:
The game and data are based on Tribe and Peacock (1976) as simplified by Betty Nicotri, University of Washington, 1986 (personal communication).

9) COMMUNICATING RESULTS AND CONCLUSIONS BASED ON EVIDENCE

It is important that your students practice communicating science. Communication involves speaking (as in a formal or informal presentation), creating a display of one's work that can be reviewed by others, or listening carefully to a presentation and being able to process what is being said (listening is a form of communication).

Another form of communication is drawing (creating illustrations of what students understand about a concept). Asking students to create a drawing or sketch about a biological phenomenon can be used in both a formative or summative way. For instance, used in a formative way, asking students to draw a DNA molecule without looking at a book, illustration, or notes and before you talk about it in class is a way for you to assess their current level of understanding or to reveal their misconceptions they have. Ask students to take out a blank piece of paper and draw, in detail, with labels, an object or a process, such as the light-dependent reactions of photosynthesis. This will allow you to accomplish two things: to get the students to see where they "get stuck" in the drawing, which is related to helping them think about their own thinking—metacognition, and to help you to determine what you need to focus on to help your students. Do they have misconceptions that need dispelling or is there one part of a process that is particularly difficult for most/all of your students? Drawings are also fairly easy to grade, so they can be reviewed relatively quickly.

A related activity is to give students a short reading—articles from Science News are often really good because they include data and experimental design and the scientist's rationale for doing an experiment. The article below is about the relationship between diet and breast cancer. Students read the article, and then are asked to create an illustration that could be used in their textbook. The illustration could include stick figures, a few words, numbers, or whatever they would like to use to COMMUNICATE what they think is the "essence" of the article. In other words, what do they think are the most important points in the article (it is important for students to decide what they think is important), and how would they illustrate those important ideas so that someone who has not yet read the article could understand the content of the article just by looking at their illustration? Get them to think about the good illustrations in their texts and how much information can be gleaned from them.

Allow students to work alone first, and then have them compare their drawings. In this article, the most important ideas can be found in the first two columns. Also, note the importance of the information about the relative concentrations of the cabbage extract (parts per million vs parts per billion; ppm vs ppb).

The following is a drawing made by one student who seems to have captured the essence of the article "*Fighting cancer from the cabbage patch*" and communicated well about her understanding of the experiments and the results.

Resource:
Fighting cancer from the cabbage patch
    http://www.sciencenews.org/pages/pdfs/data/2000/158-13/15813-09.pdf

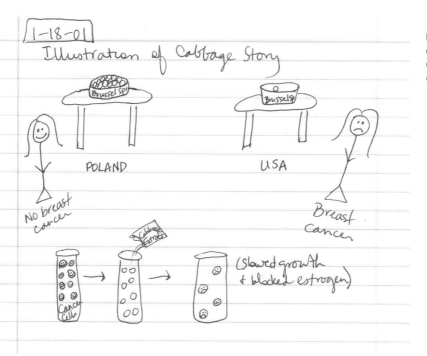

Figure 1.28 Student drawing to illustrate essence of science article.

10) CONNECTING AND RELATING CONTENT KNOWLEDGE TO OTHER CONCEPTS TO CONSTRUCT A FRAMEWORK OF UNDERSTANDING BIOLOGY

Connecting Content

One of the best ways that you can help your students to gain a deeper understanding of biology is to help them make connections between the content that you teach. If students cannot make connections between two topics in one concept or two concepts to each other, they are not seeing the "big" picture. So, you can help them scaffold the biological information they are learning, help them to gain a framework on which to build their knowledge, and to give them opportunities to make connections between content as often as possible. This also breaks them of the bad practice of memorizing content in separate silos—the more connections they make, the less they have to memorize facts and the easier it becomes for them to see how new information (to them) is related to something that they have already "learned." This is related to the idea of "constructivism" and how students build their content knowledge. Just remember.… "everything is connected in biology!"

The following is a *Science*Now article that has been used to make connections of biological content to information that has been previously discussed in class: http://news.sciencemag.org/sciencenow/2001/04/11-01.html. There would certainly be other parts of the article that could be linked to what you teach, but each number represents one idea of making a connection.

1. Connect to different types of reproduction plants exhibit—sexual reproduction and asexual reproduction. This describes a case study in pollination biology, which is a prelude to sexual reproduction in a flowering plant.

2. This is related to the misconception that many people have about the differences between plants and animals: many people think that plants only conduct "photosynthesis" while animals go through "cellular respiration" —not realizing that plants also respire. This should also be linked to the overall process of cellular respiration noting that one of its waste products is heat.

3. Organisms within a community interact. One of the ways that they interact (and co-evolve) is when plants are pollinated by animals. To establish and maintain this interaction, plants (and flowers) must first attract animals to the flower and then reward the animals so that they visit other flowers with similar characteristics, which would allow the distribution of pollen from one plant to another. Attractants are characteristics such as floral odor, movement of floral parts, color and size of the flower, while rewards include nectar, pollen (as long as all of the pollen isn't eaten!), shelter (and warmth as in this case), and even sexual gratification (investigate wasp pollination of orchids).

4. This gets at the concept of homeostasis in organisms and the requirements for energy and matter to regulate physiological characteristics, such as temperature.

5. Relate this to the differences between aerobic and anaerobic respiration, and how much heat would be generated in either of these processes. This section can also be related to the differences between plants and animals—both require oxygen for aerobic respiration, but humans have to have lungs and a circulatory system to distribute oxygen efficiently to all parts of the body. This is why people participate in "aerobic" exercise—to help their body be able to distribute oxygen efficiently.

6. Talk about where students have seen stomata before (in leaves), and the function of the stomata there (allow gas exchange). Also, if stomata are found on the petals of plants, what does this indicate

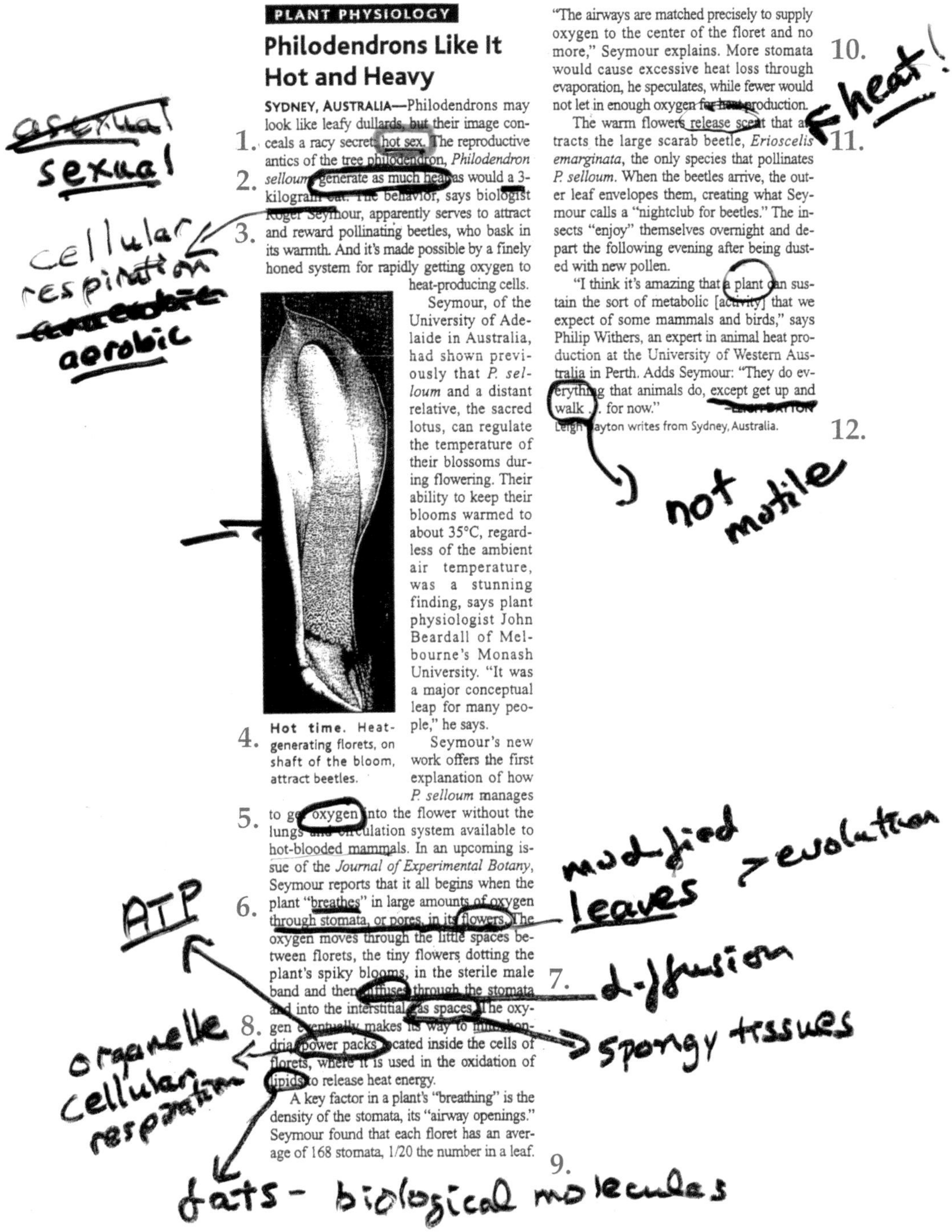

Philodendrons Like It Hot and Heavy

SYDNEY, AUSTRALIA—Philodendrons may look like leafy dullards, but their image conceals a racy secret: hot sex. The reproductive antics of the tree philodendron, *Philodendron selloum*, generate as much heat as would a 3-kilogram cat. The behavior, says biologist Roger Seymour, apparently serves to attract and reward pollinating beetles, who bask in its warmth. And it's made possible by a finely honed system for rapidly getting oxygen to heat-producing cells.

Seymour, of the University of Adelaide in Australia, had shown previously that *P. selloum* and a distant relative, the sacred lotus, can regulate the temperature of their blossoms during flowering. Their ability to keep their blooms warmed to about 35°C, regardless of the ambient air temperature, was a stunning finding, says plant physiologist John Beardall of Melbourne's Monash University. "It was a major conceptual leap for many people," he says.

Hot time. Heat-generating florets, on shaft of the bloom, attract beetles.

Seymour's new work offers the first explanation of how *P. selloum* manages to get oxygen into the flower without the lungs and circulation system available to hot-blooded mammals. In an upcoming issue of the *Journal of Experimental Botany*, Seymour reports that it all begins when the plant "breathes" in large amounts of oxygen through stomata, or pores, in its flowers. The oxygen moves through the little spaces between florets, the tiny flowers dotting the plant's spiky blooms, in the sterile male band and then diffuses through the stomata and into the interstitial gas spaces. The oxygen eventually makes its way to mitochondria "power packs" located inside the cells of florets, where it is used in the oxidation of lipids to release heat energy.

A key factor in a plant's "breathing" is the density of the stomata, its "airway openings." Seymour found that each floret has an average of 168 stomata, 1/20 the number in a leaf.

"The airways are matched precisely to supply oxygen to the center of the floret and no more," Seymour explains. More stomata would cause excessive heat loss through evaporation, he speculates, while fewer would not let in enough oxygen for heat production.

The warm flowers release scent that attracts the large scarab beetle, *Erioscelis emarginata*, the only species that pollinates *P. selloum*. When the beetles arrive, the outer leaf envelopes them, creating what Seymour calls a "nightclub for beetles." The insects "enjoy" themselves overnight and depart the following evening after being dusted with new pollen.

"I think it's amazing that a plant can sustain the sort of metabolic [activity] that we expect of some mammals and birds," says Philip Withers, an expert in animal heat production at the University of Western Australia in Perth. Adds Seymour: "They do everything that animals do, except get up and walk . . . for now."

—LEIGH DAYTON

Leigh Dayton writes from Sydney, Australia.

Handwritten annotations:

1. ~~asexual~~ sexual
2. cellular respiration ~~anaerobic~~ aerobic
3. (arrow to "heat-producing cells")
4. (arrow to image)
5. oxygen
6. ATP — organelle cellular respiration
7. diffusion → spongy tissues
8. modified leaves → evolution
9. fats — biological molecules
10. ← heat!
11. (circle around "metabolic activity")
12. not motile

about the evolutionary history of flowers and petals (that petals evolved from leaves)?

7. A discussion of interstitial gas spaces can be used to remind students of tissues in other parts of plants and other organisms. In plants, "spongy tissues" can be found in different locations that facilitate the free exchange of gases with the environment. One example is found in many leaves, where the spongy parenchyma next to stomata in the bottom half of leaves allow for better gas exchange between the outside air and the palisade parenchyma in the upper part of the same leaves. This arrangement of cells allows carbon dioxide to enter through the stomata and get to both the spongy parenchyma and palisade parenchyma in the leaves.

8. Mitochondria are the power packs of the cells, which can remind students about the different organelles a plant cell possesses, and their functions.

9. Students might need help with the concept of "oxidation," however, another connection can be made when you bring up the idea of how biological molecules are made and used within a plant. The biological molecules are fats, proteins, carbohydrates, and nucleic acids. When an organism, such as a plant, takes in or produces a lot of energy-containing materials (like sugar), some of that energy can be stored as biological molecules, such as fat. However, when a plant, or an animal, is not taking in energy (during the winter for plants or when an animal can't find food), their body still requires ATP. So, these organisms must break down their own stored biological molecules, such as fat, and use them to produce the ATP their cells constantly require to remain alive.

10. "Tradeoffs" is a biological concept that applies to many different structures, situations, and processes in living organisms. A tradeoff identifies both the positive and negative aspects of the same characteristic. Here, having more stomata would be good because then more oxygen could diffuse into a philodendron during flowering, however, as noted, more stomata would also mean that excessive heat would be lost through these holes. On the other hand, fewer stomata would probably not allow enough oxygen into the flowers to support the high rate of aerobic respiration and generate the amount of heat---also part of the tradeoff.

11. Here, you can connect this information to the production of heat mentioned in #2—of what value is the production of heat to philodendrons? It's not just the heat that is used by the beetles to warm themselves (a reward), but the heat helps disperse a scent

that attracts the beetles to the flowers in the first place (you have to attract animals to the plant before they can pick up pollen and carry the pollen to other philodendrons)

12. Again, what are the main differences between plants and animals? Animals are generally motile, while plants are not. This has major consequences for the lives of these organisms. For instance, while animals can move around seeking mates for sexual reproduction, plants cannot. Therefore, they require a pollination agent (such as the wind or animals) to carry their sperm (inside pollen grains) to other plants so that sexual reproduction can occur. The fact that plants are non-motile has many consequences related to their growth and reproduction.

Constructing a Framework of Understanding

Biology is often compared to a foreign language because, in many cases, the course focuses on terms and vocabulary words. It is difficult to avoid at least some terminology—regardless of your attempt to streamline your course so that you emphasize processes, major concepts, and understanding instead of just vocabulary. Your biology course will require some vocabulary. Often, however, students are unable to distinguish between the importance and the meaning of a biological word. For instance, when studying photosynthesis, are the terms "chloroplast" and "thylakoid" of equal importance to the understanding of the process of photosynthesis? Are both of these words necessary for the understanding of photosynthesis? Also, what is the difference between the nature of the words "chloroplast" and "Calvin Cycle?" Students may simply memorize definitions of words without gaining an understanding of what they mean or their significance or how the terms fit together.

One way to help students think about the difference in the meaning and importance of biological terms is to give them a list of those terms you have used after several lessons on related concepts and get them to organize these terms in some meaningful way. For instance, the following list of terms all have something to do with plants—but some of the terms are structures and some of them are processes.

Have students organize these words so that they are arranged in small groups. To ensure that there is understanding of the groupings and the terms, have students assign a heading (a word or two) that describes the words that are grouped in that particular collection of terms. It is important to note that there is no single correct way to group these terms, however, each grouping must make sense biologically. Also, forming a single large group of words is not a valid way to organize the words, although finding a title for the entire word set would be important (and not necessarily easy).

Organize Terms into Logical Groups with Headings and Subheadings

- Net venation
- Axillary bud
- Parallel venation
- Simple Blade
- Node
- Compound Leaf
- Dicot Leaves
- Tendril
- Sheath
- Spine
- Internode
- Simple Leaf
- Leaflets
- Transpiration
- Opposite
- Cotyledons
- Alternate
- Bud scales
- Monocot Leaves
- Whorled
- Petiole
- Stipules
- Photosynthesis

INSTRUCTIONS FOR STUDENTS:

All words must be placed into a group.

Each word can be placed in only one group.

Provide a heading (for major groups) or subheading (for subgroups) for each of your groups.

No major group can contain only one term (although subgroups may).

You must have at least two different groups of words, however, you may have many more than two groups (but not more than 11).

Write a one-sentence explanation of why you arranged the words in each major group.

Note: There are many possible ways to organize these words and to describe the different groups. Here is one possible way:

Leaf Structures and Functions
Leaf Functions
       Transpiration
       Photosynthesis

Arrangement of Leaves
    Opposite
    Alternate
    Whorled

Plant Structures Associated with Leaves and Their Attachment to Stem
    Axillary Bud
    Node
    Internode
    Stipules

Characteristics Used to Distinguish Leaves
    Monocot Leaves
        Sheath
        Parallel Venation
    Dicot Leaves
        Net venation
        Petiole
    Type of Leaf
        Simple Leaf
            Simple Blade
        Compound Leaf
            Leaflets

Types of Modified Leaves
    Tendril
    Spine
    Bud Scales
    Cotyledons

2 Linking Content And Practices To Improve Student Learning

2.1. Dealing with Content

Now that you've considered the student attitudes you want to promote and the processes you want students to understand and be comfortable with, it's time to consider the content you want to help students learn. Content is not a commodity that you can measure out and pour into your students' heads. It is not something you can "cover" in your class and be confident that your students know and understand automatically. In fact, there is now a growing body of evidence that supports the old adage "less is more." Even the late Neil Campbell, whose encyclopedic textbook continues to set the standard for AP and introductory college biology courses, was adamant that he did not expect teachers to "cover" the entire book, but rather he was providing a resource that could be used to go into greater depth on any specific topic of the teacher's choosing.

Your first important task is to choose the major concepts and topics on which you should focus. Most of this work has already been done for you if you work in the public schools. Every state has devised state standards for content instruction at different grade levels. For the most part, these are based on the National Science Standards, but there is considerable variation from state to state. The appropriate state guidelines should be your primary guide for selecting those concepts on which to focus. Another good model is the new AP Biology Curriculum Framework (2011). A working group of plant scientists recently outlined what undergraduate biology majors should learn about plants and organized concepts using the life science domains in the Next Generation Science Standards (http://c.ymcdn.com/sites/aspb.site-ym.com/resource/resmgr/Education/Undergradplantbio_conceptsan.pdf). These efforts to vertically integrate progression across K-16 levels help address college readiness and support the focus on big ideas, core competencies, and student-centered learning. In addition, the American Society of Plant Biologists provides a tool for integrating plants into the National Science Standards http://my.aspb.org/?page=EF_Principles

For each concept you want your students to understand, you will have to decide on one or more illustrative examples. Of course, we suggest using a plant example whenever possible. Begin by finding an image, a data set, or a question that you can use to lead students into thinking about the concept you are covering. Use student answers to focus further questions to guide their thinking. The botanical educator William Beal expressed the key—and timeless—principles of inquiry way back in 1880:

> "The pupils are not told what they can see for themselves . . . In the whole course in botany I keep constantly in view how best to prepare students to acquire information for themselves with readiness and accuracy."

Our task as teachers is to help students see things for themselves; to create their own understanding of the concept built upon their prior knowledge. Success in guiding inquiry depends not only on your prior preparation to set up the learning situation, but on your interactions with students during the class. Whenever possible, avoid directly answering questions they raise. Instead, try to return a question that will lead them to the answer. For instance, in the Celery Challenge activity (Unit 3) a student may ask if it will make a difference how long they cut their celery pieces. Instead of answering directly, reply "It might, how would you find out?" We also find it useful at the "end" of an activity to challenge students to think if there might be an alternative explanation for their results. A good example of how to do this is in the Phototropism activity with Darwin's experiments. Especially in the lab, we tend to train students to think that when they get a result, their job is done. Their results will either support what they expected to happen, or, if not, the explanation is some "experimental error" in how they performed the experiment. This is not how science works. We do an experiment to answer a question, but the answer always leads to more questions.

The Ecology Game (Chapter 1), Carbon Cycle (this Chapter), and Climate Change (Chapter 3) activities are seemingly straight forward at first, and students will expect a certain answer. However, the data provide some unanticipated results. The data in the latter two activities clearly suggest that there must be an alternative explanation. This forces students to look beyond the "obvious." The Carbon Cycle activity is particularly useful because the observed discrepancies could superficially be interpreted as "experimental error." It provides an opportunity to teach students to accept their data, even if it is not exactly what they expected, and search for alternative explanations based on their results.

In 1986 Ken Costenson and Anton Lawson published an article called "Why isn't inquiry used in more classrooms?" The second most cited reason was that inquiry was too slow, and "We have district curricula and must cover all the material." The implication is that the only way students can learn any material is if it is "covered."

In fact, the last 20 years of research demonstrates that students who focus, with more depth of understanding, on fewer concepts perform better than traditionally taught students on content assessments and high-stakes exams—even on topics that were not specifically "covered." Inquiry is one of the most useful tools for teachers to use to help their students develop such depth of understanding. Some of the relevant literature is included below.

Following the bibliography are three activities as examples of ways to focus on the big ideas. The chapter then shifts from teaching to learning.

A BRIEF BIBLIOGRAPHY OF RESEARCH STUDIES DEMONSTRATING THAT FOCUSING ON BIG IDEAS AND MAJOR CONCEPTS IS MORE EFFECTIVE THAN "COVERING" CONTENT IN PROMOTING STUDENT LEARNING.

Andrews, T.M., M.J Leonard, C.A. Colgrove, and S.T. Kalinowski. 2011. Active learning *NOT* associated with student learning in a random sample of college biology courses. *CBE-Life Sciences Education* 10: 394–405.

Armbruster, Peter, Maya Patel, Erika Johnson, and Martha Weiss. 2009. Active learning and student-centered pedagogy improve student attitudes and performance in introductory biology. *CBE-Life Sciences Education* 8: 203–213.

Belzer, Sharolyn, Micha Miller, and Stephen Shoemake. 2003. Concepts in Biology: A supplemental study skills course designed to improve introductory students' skills for learning biology. *American Biology Teacher* 65(1): 30–40.

Brickman, Peggy, Cara Gormally, Norris Armstrong, and Brittan Hallar. 2009. Effects of inquiry-based learning on students' science literacy skills and confidence. *International Journal for the Scholarship of Teaching and Learning* 3: 1–22.

Coil, David, Mary Pat Wenderoth, Matthew Cunningham, and Clarissa Dirks. 2010. Teaching the process of science: Faculty perceptions and an effective methodology. *CBE-Life Sciences Education* 9: 524–535.

Dirks, Clarissa and Matthew Cunningham. 2006. Enhancing diversity in science: Is teaching science process skills the answer? *CBE-Life Sciences Education* 5: 218–226.

Freeman, Scott, Eileen O'Connor, John W. Parks, Matthew Cunningham, David Hurley, David Haak, Clarissa Dirks, and Mary Pat Wenderoth. 2007. Prescribed active learning increases performance in introductory biology. *CBE-Life Sciences Education* 6: 132–139.

Freeman, Scott, David Haak, and Mary Pat Wenderoth. 2011. Increased course structure improves performance in introductory biology. *CBE-Life Sciences Education* 10: 175–186.

Kitchen, Elizabeth, Joan D. Bell, Suzanne Reeve, Richard R. Sudweeks, and William S. Bradshaw. 2003. Teaching cell biology in the large-enrollment classroom: Methods to promote analytical thinking and assessment of their effectiveness. *Cell Biology Education* 2: 180–194.

Klionsky, Daniel J. 2002. Constructing knowledge in the lecture hall: A quiz-based, group-learning approach to introductory biology. *Journal of College Science Teaching* 31(4): 246–251.

Knight, Jennifer K. and William B. Wood. 2005. Teaching more by lecturing less. *Cell Biology Education* 4: 298–310.

Lord, Thomas R. 1999. A comparison between traditional and constructivist teaching in environmental science. *The Journal of Environmental Education* 30: 22–28.

Moravec, Marin, Adrienne Williams, Wancy Aguilar-Roca, and Diane K. O'Dowd. 2010. Learn before lecture: A strategy that improves learning outcomes in a large introductory biology class. *CBE-Life Sciences Education* 9: 473–481.

Preszler, Ralph. 2004. Cooperative concept mapping: Improving performance in undergraduate biology. *Journal of College Science Teaching* 33: 30–35.

Sundberg, M.D. and M.L. Dini. 1993. Majors vs nonmajors: Is there a difference? *Journal of College Science Teaching* 22: 299–304.

Sundberg, M.D., M.L. Dini, and E. Li. 1994. Improving student comprehension and attitudes in freshman biology by decreasing course content. *Journal of Research in Science Teaching* 31: 679–693.

Sundberg, M.D. and Gregory G. Moncada. 1994. Creating effective investigative laboratories for undergraduates. *BioScience* 44: 698–704.

Sundberg, M.D. 2003. Strategies to Help Students Change Naive Alternative Conceptions about Evolution and Natural Selection. *Reports of the National Center for Science Education* 23(2): 23–26.

Travis, Holly and Thomas Lord. 2004. Traditional and constructivist teaching techniques: Comparing two groups of undergraduate nonscience majors in a biology lab. *Journal of College Science Teaching* 33: 12–18.

HOW TO FOCUS ON THE BIG IDEAS

Plant Growth and Physiology

In a nutshell: This activity focuses on the big ideas about the structures of plants that enable their life functions and how they grow and develop. Through a series of images and questions, students consider functional adaptations of root form and growth and connect understandings of cell division, cell elongation, cell differentiation, osmosis, and gravitropism.

In the examples below, roman font provides a model of background information and questions you can use to lead student thinking. Indented italicized script illustrates typical student response.

•→ Figure 2.1 is a photograph of growing root tip seen through a dissecting microscope. Describe in words the appearance of this organ as you move from the tip (bottom) of the root towards the base (top).

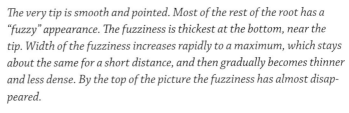

> *The very tip is smooth and pointed. Most of the rest of the root has a "fuzzy" appearance. The fuzziness is thickest at the bottom, near the tip. Width of the fuzziness increases rapidly to a maximum, which stays about the same for a short distance, and then gradually becomes thinner and less dense. By the top of the picture the fuzziness has almost disappeared.*

•→ The tip is smooth and pointed. How might this shape be a functional adaptation for the way roots grow?

> *The root has to push its way through the soil, so being smooth and pointed, like a nail, would make it easier to penetrate through soil.*

•→ The fuzziness is caused by thin extensions of a single cell, growing sideways out of the surface of root, called root hairs. How might this be a functional adaptation related to the requirements a plant needs for its growth?

> *Root hairs help the root absorb water and nutrients.*

Figure 2.1 Photograph of a living root tip viewed with light microscopy.

•→ The thousands of tiny root hairs, growing between soil particles, do increase the surface area of the root that allows for more uptake of water and minerals from the soil. When thinking about roots, the first thing that likely comes to mind is the important role in absorbing nutrients that are carried to the rest of the plant and water the entire plant needs for basic life functions. Roots have additional adaptations. Water uptake is especially important for root growth because this will provide the pressure to push the tip through the soil. However, most roots have an even more effective way of doing this. Approximately 80% of all flowering plants and virtually all gymnosperms, ferns and other vascular plants have symbiotic associations with filamentous fungi, mycorrhizae, whose wide network of cells absorb water and nutrients for the host plant. If you're on a trip and want to collect seeds of an interesting plant to grow at home, be sure to collect some of the soil from the area, along with the seeds. This way you'll be able to inoculate your potting soil with fungi that will establish the mycorrhizal association with your seedlings and promote their growth.

There is another equally important function of root hairs for root growth. Here's a hint: think about what the roots look like when you pull a weed from the soil.

Inquiring About Plants

Roots anchor the plant to the soil, and roots hairs, which grow in between individual soil particles, have a role in making this a firm anchorage. As a result, when you pull up a weed, soil particles remain attached to the roots.

This anchorage helps the plant resist uprooting in a storm and also helps hold the soil in place to prevent erosion. However, we're interested in how this same anchorage helps the root grow. As the root tip pushes into the soil, what keeps this force from pushing the plant out of the soil?

Growth by the cells between the root tip and the root hair region creates pressure in both directions. Anchorage by the root hairs provides more resistance against this force than the soil particles in front of the root tip. The root hairs provide a firm anchor against which the growing root tip can push.

An interesting modification occurs in many bulbs and corms, such as onion, crocus, and lilies, that produce "contractile roots." In these plants, the bulb can be pulled deeper into the soil by specialized roots. The root hair region is necessary to anchor the root before beginning contraction.

Now let's focus on the smooth pointed tip of the root. Here is a microscopic view of part of a longitudinal section of a corn root tip. Corn has large roots, which makes them easy to work with, and there is a distinct boundary between the root axis and the root cap (arrow). Given the appearance of cells in this photo, where in these roots do you think cell division is centered? What is your evidence?

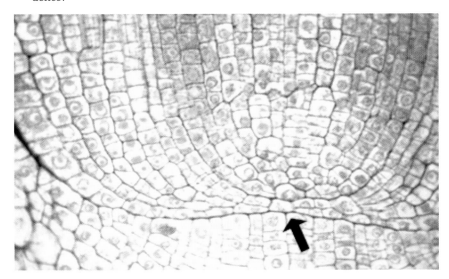

Figure 2.2 Microscopic view of corn root tip section.

The center of growth seems to be right at the tip of the arrow. Files (rows) of cells all seem to come from this region.

→ If you ignore the root cap at the bottom of the slide, the cells of the root appear to be arranged in files (rows) that flow from a point just above the tip of the arrow. How would the cells have to divide to form a single file?

To form a file of cells, every time a cell divides, the new cell wall must be perpendicular to the axis of the file.

→ The bottom file of cells, above the distinct border with the root cap, seems to converge on a single cell above the arrowhead. On either side of this cell are two stacked cells. What must have happened if a single cell gave rise to the stacked pair of cells on the left?

First, the cell must have divided normally with the new wall between daughter cells at right angles to the file. Then, the daughter cell on the left must have divided again, but this time with the new wall parallel to the axis.

→ The pair of cells on each side of the central cell of the lower file has a combined thickness similar to the thickness of the central cell. This suggests that their respective mother cells were about the same size as the single central cell and that the mother cells must have divided recently. This mechanism of increasing the number of files occurs frequently near the tip. For instance, to the left of the first pair are four cells in two pairs and two more splits are clearly seen from the second file of cells. This pattern of cells suggests that there must be active cell division in this region.

Do you see any other evidence of cell division in this region?

A few cells appear to be in various stages of mitosis, although it is difficult to tell at this magnification and resolution. The cells are mostly small and square or in pairs equal size.

→ In a few files of cells, near the center, the cytoplasm stains more densely. What might this suggest?

Denser staining suggests more organelles and higher metabolism, either for rapid cell division or for differentiation into different cell types.

→ The patterns you've observed were also seen by botanists in the late 1800s. Based on these observations, this region was named the root apical meristem. Meristem is defined as a region of actively dividing cells producing stem cells that can differentiate into specialized cell types. Near the tip, apex, of every growing root is a root apical meristem.

Figure 2.3 is a diagram of a growing root tip that shows the position of the root apical meristem, just behind the root cap and in front of the root hair zone. For almost 100 years it seemed perfectly obvious that mitosis must be centered in the apical meristem. Then, in 1956, an Englishman named Clowes realized that no one had ever tested this hypothesis because it seemed so obvious. How could you test this hypothesis in the laboratory right now?

You could take a microscope slide of a root tip and count the number of cells in mitosis in the center of the apical meristem.

→ Clowes made microscope slides of a number of root tips and counted the number of mitotic figures in the center of the apical meristem. To his surprise, there were almost NONE in the very center! This was a surprising discovery that went against the "obvious" explanation, so he decided he'd better find another way to test it. By then, radioactive isotopes, including tritiated thymidine, were available that could be used to study biological processes. Thymidine is one of four nucleosides in DNA, and tritium is a radioactive isotope of hydrogen. Clowes grew some roots in a solution of tritiated thymidine so that every time a cell divided the new nucleus became slightly radioactive. He then sectioned the roots and mounted them on microscope slides. But before putting on cover glasses, he went to the dark room and dipped the slides into melted photographic emulsion. He left them to dry in the dark overnight. During that time, the radioactivity exposed the film over every radioactive nucleus. The next day he developed the slides exactly the same way photographic film is developed, and the slide,

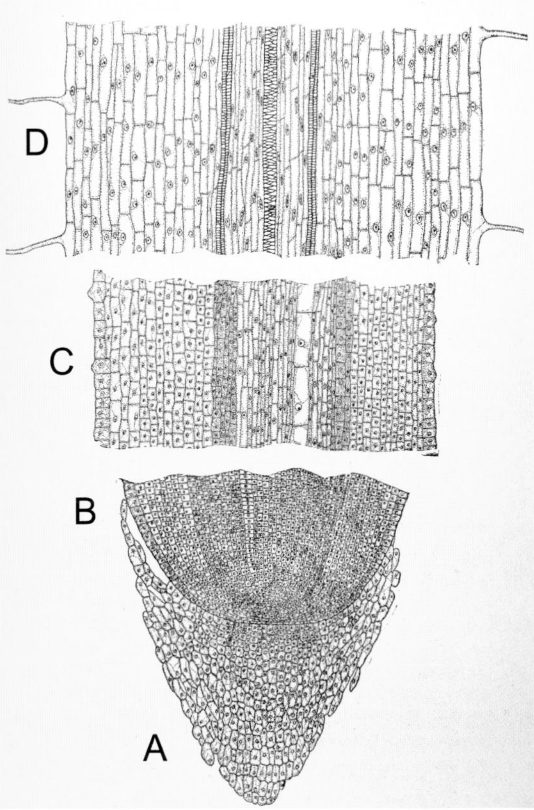

Figure 2.3 Diagram of a longitudinal section through a corn root tip: A. Root cap. B. Region of cell division. C. Region of elongation; D. Region of maturation.

thus, became a film negative. Every place the film was exposed there was a black dot. Below is an example of what he saw (Figure 2.4).

→ How many cells divided overnight in the center of the root apical meristem?

No cells divided in the center of the meristem.

→ Now Clowes had two lines of evidence that demonstrated that the center of the apical meristem, where we thought cell division should be most active, actually had little or no division. He called this region the "quiescent center"—the center of the apical meristem was mitotically quiet. We now know that the quiescent center is a reserve of cells that become activated if the root tip is damaged. In the same way that physical injury stimulates our cells to begin to divide as a wound response, injury stimulates the quiescent center cells to divide to repair damage to the root tip.

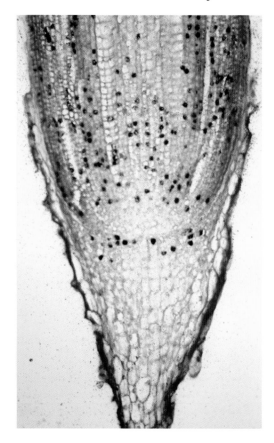

Figure 2.4 Auto-radiograph of an onion root tip. Black dots are nuclei that divided during the 24 hours the roots were grown in a solution containing radioactive thymidine.

Adding new cells, around the periphery of the apical meristem, will help to force the tip further into the soil. Look back at the diagram of a longitudinal section of a root. What evidence is there of another mechanism that generates a pushing force?

Cells get longer in the zone of elongation.

▸▸ The smallest cells occur in the meristem where frequent mitosis divides mother cells into two smaller daughter cells. The longest cells occur in the region where root hairs are formed and functioning. Elongation occurs between these regions, driving the tip forward against the resistance provided by the anchored mature cells in the root hair region.

What can provide the force required to elongate the cells and push the root deeper into the soil?

Osmosis can move more water into the cell's vacuole stretching it, which in turn stretches the entire cell.
New cell wall can be formed in one direction only.

▸▸ Plants produce hydraulic pressure as a result of osmosis in the developing cells produced by the apical meristem. Osmosis is the driving force, but physical resistance is the limiting factor.

At the level of a single cell. Consider an animal cell bound only by a cell membrane, for instance, a red blood cell (unfortunately not as readily available in the high school or college lab as they used to be). What do you think will happen to that cell if you put it into distilled water?

The cell will swell up as water moves in.

▸▸ As water diffuses across the membrane by osmosis, from the region of higher free energy of water (outside the cell) to the region of lower free energy (the cytoplasm), the cell swells, the membrane stretches and starts to leak, and eventually the cell bursts and the cytoplasm flows out of the cell into solution.

Now consider the plant cell where a similar membrane is bounded to the outside by a semi-rigid cell wall. The first thing to realize is that the cell wall is NOT solid! Rather, the cellulose is like a plastic mesh bag sometimes used for packag-

Inquiring About Plants

ing fruits. The mesh itself is very flexible, but it has a lot of tensile strength (it cannot be pulled apart very easily; Fig 2.5A).

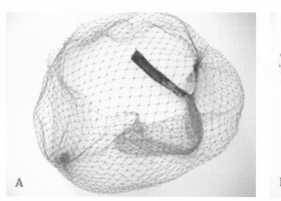
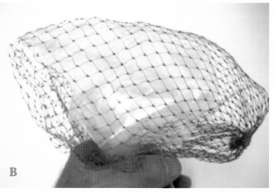

A   B

•▸ The mesh-like cell wall, which was produced by the cell membrane, surrounds the membrane the way the clear plastic bag above is inside the mesh bag (Fig 2.5B).

Now, if the plant cell was placed into the same distilled water as the animal cell was, what do you expect would happen?

The cell will still swell, but the membrane will keep it from bursting.

•▸ The cell would still take up water and swell, but it would be restricted by the outer mesh wall, which would firmly resist swelling (Figure 2.6).

Figure 2.5 Model of a plant cell. A. Cell wall. B. Cell membrane and cell wall.

Figure 2.6 Model of a plant cell wall and membrane "stretched" due to osmosis.

•▸ Note that in the model above, the mesh is stretched tight but the individual plastic filaments do not stretch—cellulose does basically the same thing in the plant cell wall. By orienting how the original cellulose mesh is laid down, the cell can regulate which way it might "stretch" when enough water is taken in to force the cell membrane against the mesh-like cellulose wall. Of course, the more water pressure that builds up inside the cell, due to osmosis, the more forceful the cellulose walls resist stretching and the firmer the cell becomes. We say the cell is turgid because the resistance pressure of the cell wall is equal to the expanding force of water pressure in the vacuole. If most of the reinforcing cellulose is in bands around the sides of the cell, the cell wall will be forced to elongate rather than simply enlarge in all directions.

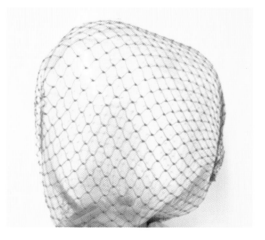

Teaching Plant Anatomy

In a nutshell: This activity is a model for teaching any kind of structural information and can be modified for teaching structure/function relationships.

↦ Examine the image below of the stem of maize (*Zea mays*, a monocot) and make a sketch of the general patterns you see. Make your sketch large enough to fill an entire sheet of paper. Do not worry about putting in any cellular details at this point. All you want to do is indicate any observable patterns.

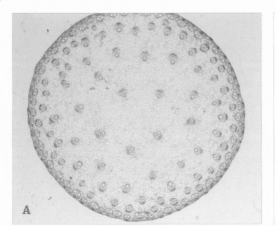
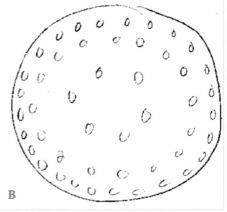

Figure 2.7 A. Cross section of maize stem. B. Typical student sketch of maize stem.

Circulate through the class with a blank overhead and pen. When you find a student with a sketch similar to the one above, ask him/her to quickly copy the sketch onto the overhead to use as an example. This is the sketch you will label and describe as the class discusses the structure they see. Most of you will have a sketch similar to the one shown above.

↦ The line around the outside represents the epidermis, a layer of tightly packed cells that secretes a waxy substance (if aboveground). In many respects this layer is similar to the skin on your body. What might be a function of the epidermis in stems?

Some kind of protection is usually brought up. Expand from whatever the first student brings up. "Protection" has many aspects: 1) from water loss, 2) from microbes, especially a variety of pathogens, 3) from physical injury. The epidermis also provides some physical strength to support the structure. Think of a long balloon. The latex wall is not very strong, but it does provide a boundary between the inside and outside. In a sense it protects the balloon from losing air (analogous to water loss from the stem) when it is blown up. The filled balloon is rigid because the wall resists internal air pressure. In the same way, part of why an herbaceous stem is rigid is because the epidermis resists the internal water pressure of all the internal cells that cannot lose their water by evaporation from the surface.

↬ The circular structures inside the stem are vascular bundles. We will look at them more closely in a minute, but first, look carefully at how they are arranged. Textbooks usually say that the bundles in monocot stems are "scattered" or "embedded throughout." However, several patterns are apparent in your diagrams. What is one of the patterns?

> *The two patterns usually mentioned first are that there are more vascular bundles near the epidermis and that these bundles are smaller than those nearer the center of the stem. A third pattern that is usually mentioned is that the bundles appear to be in concentric rings. Occasionally a student will notice that the orientation of bundles (xylem vs phloem) is always the same. We usually look at this separately with a close up of a few of the bundles, then come back to this slide to "test" their hypothesis about arrangement.*

↬ The patterns you noticed in this image went largely unnoticed by botanists until the 1970s. We now know that the bundles in monocot stems are very complexly, but very predictably, arranged in the stem. In fact, every one of the bundles in this image will either unite with another one or enter a leaf somewhere further up the stem. An expert could predict where every single bundle will go—where it will fuse or which vein of which leaf it will become!

Figure 2.8
A. Magnified view of maize stem.
B. Sketch of magnified view of maize stem.

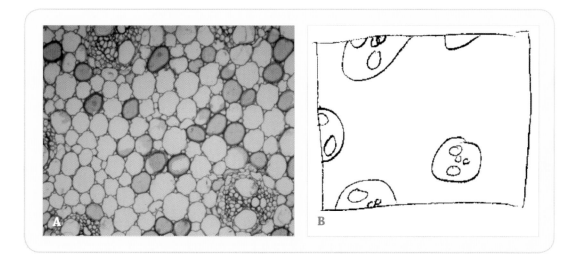

All of the rest of the tissue in this stem is simply called ground tissue and is made up of large, thin-walled cells containing mostly central vacuole.

Figure 2.8 is a close-up of a portion of the inner part of the previous slide. Again, make a sketch of the general patterns you see. Don't worry about individual cells yet.

Again, go around the class with a clean overhead and ask a student to copy his/her sketch to use as the class example. Many of you have something that looks like the above sketch.

•→ What are the mostly circular areas containing smaller cells?

These are the vascular bundles.

•→ These are the vascular bundles and you can see a very regular arrangement of cells within them. They are dominated by two large openings that are about the same size as the ground tissue cells between the bundles. Sometimes you can see pieces of stuff inside these openings. These are the remains of the original xylem cells that used to be in these positions. This was the first-formed xylem when the stem was much smaller. As development occurred and new cells formed, the original xylem cells were stretched until they broke apart. These opening are holes (lacunae) where the original xylem was located.

To some people these bundles look like a monkey face or a Halloween mask. For instance, if the complete bundle near the bottom of the image was a mask, the two xylem holes would be the "eyes." What appear to be the "mouth and nose" are the functioning xylem in these bundles. They have evenly thick walls and are empty inside. In the "forehead" region is a distinct patch of smaller cells. These are the phloem cells. Xylem and phloem are in the same bundle in roots.

In the ground tissue there appears to be two types of cells. Take another look at the image and make a sketch of a group of 4 or 5 connected cells including both types.

Figure 2.9 Typical student sketch of portion of magnified view of maize stem.

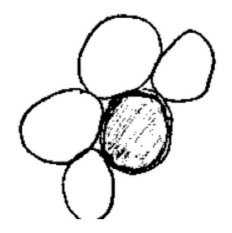

•→ What do you see wherever three cells come together?

There is a triangle between every three cells.

•→ What looks like a triangular opening between cells is an intercellular space. The individual cells of ground tissue are surrounded by air spaces. This will be true for roots and leaves as well as stems. What could be a function of these spaces? Talk this over for a minute with your neighbor. (Hint: these cells are living—what do they need to stay alive?)

The usual answer will be food or water so you'll have to do some "fishing." Ask the class to hold their breath. After a time, everyone will be breathing again even though you didn't tell them to take a breath—they're still supposed to be holding their breath. Why did they stop?

All living cells undergo respiration and ground tissue is made up of living cells. Intercellular spaces allow for gas exchange by living cells throughout the plant.
How do you explain the appearance of the two types of cells in your sketch? Talk this over.

The most common answer is that the darker cells are storing something in their vacuoles while the lighter cells are not.

It is possible that some cells are storing material in their central vacuoles and others are not. This is what we would see in the cells of stems or leaves with red or pink color patterns. However, that is probably not the case here. What is the color of the "shaded" area compared to the normal cell walls? Hint: cells are three dimensional.

The colors are both the same color of green—some of the cells must be cut through a top or bottom wall.

Microscope slides of prepared sections are usually thinner than the thickness of a single cell. In a mass of tissue, most sections will cut somewhere through the middle of the thickness of a cell, but occasionally the section will include either a top or bottom wall.
One last question on the monocot stem. On the close-up image above, which direction will be the closest to the epidermis, toward the top, the right, the bottom or the left? Use your sketch for orientation and we'll look again at the complete section. Take a minute to talk it over with your neighbor.

The closest epidermis must be to the left.

The final pattern we can observe on the monocot section is that every bundle is oriented so that the xylem is towards the center of the stem and the phloem is toward the epidermis. This will always be true in stem bundles.

Use the same format to examine and analyze the dicot stem and both monocot and dicot roots.

Figure 2.10A is a photo of a cross-section through a dicot (*Coleus*) stem. It is unusual because it is square with corners instead of round, but the pattern of tissues is typical. Make a sketch of the general patterns you see.

A typical student sketch is shown in 2.10B.

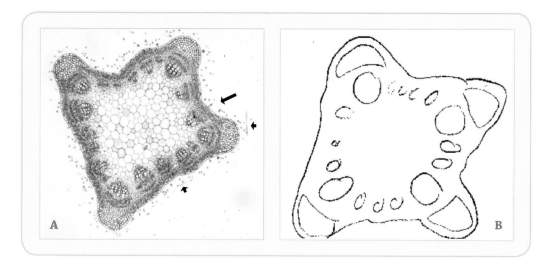

Figure 2.10
A. Cross-section
of dicot stem.
B. Sketch of
cross-section of
dicot stem.

❧ Around the outside is an epidermis, which is quite distinct in this image. What do you think the little "dots" are around the stem?

> *Students often think of bacteria or just "garbage." Occasionally a student may suggest they are hairs. A hint is to ask how the "dots" might be related to the elongate structures indicated by the shorter arrows, and the structure indicated by the longer arrow.*

❧ The "dots" around the outside are not artifacts, they are sections through individual epidermal hair cells. In a few places they are more elongate where they were cut horizontally (short arrows). A few are attached to the epidermal layer – the best example is along the upper right side (long arrow).

Does the pattern of cells at the center of the stem look anything like we saw in the roots?

> *These cells are big and empty like the ground tissue in the root.*

❧ Ground tissue in the very center is made of the same type of cells we saw in the monocot, but here they form a distinct pith, surrounded by a ring of vascular bundles. The cells of the ground tissue get smaller toward the outside, and they continue between the bundles and in a thin layer between the bundles and the epidermis. Outside the ring of bundles they form a tissue called cortex.

Figure 2.11A is a close-up of a single bundle. Sketch the general pattern that you see.

❧ On the bottom side of this bundle are distinct rows of cells. Look carefully at an individual cell from one of these rows. Does it remind you of any of the cells you saw in the monocot?

Inquiring About Plants

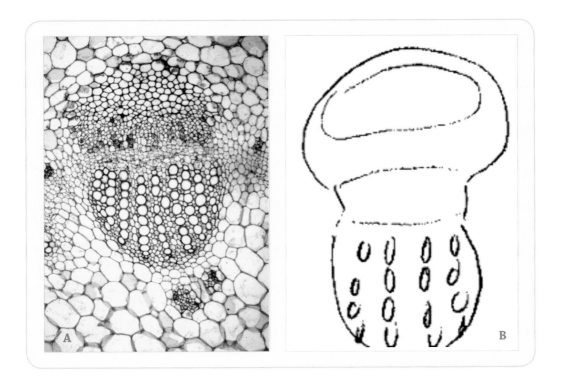

The cells have a thick round wall and look empty like the xylem cells in a root bundle.

◆▸ These are xylem cells with thick round walls and empty insides just like the xylem of the monocot. The top of the bundle contains phloem cells. How does the orientation of xylem and phloem in a dicot bundle compare to that in a monocot bundle?

In both, the xylem is toward the inside of the stem and the phloem is toward the epidermis.

◆▸ Describe the general appearance of the cells in the layer between the xylem and phloem.

These cells are smaller, flattened and have thin walls.

◆▸ These cells are a very early stage of vascular cambium that will eventually begin woody growth. In size and shape they are very similar to the cells at the bottom of the root apical meristem in Fig. 2.4 and as they divide they will form files of cells parallel to the files of xylem cells.

Finally, refer back to your sketch of the entire stem cross section (Fig. 2.7).

Figure 2.11
A. Magnified view of a single vascular bundle. B. Typical student sketch.

Beneath the epidermis in each corner of this stem are conspicuous groups of cells. Describe the appearance of these cells. Are they similar to any of the other cells you've already described?

These cells are round with thick walls and appear empty. They are similar to the xylem cells but the walls are not as thick and do not stain the same color.

→ The groups of cells in the corners, called collenchyma, have one function similar to xylem: they provide support. The walls are thick but are made entirely of cellulose whereas xylem cell walls also contain a much stiffer material, lignin.

Why do you think these strengthening cells are located in the corners? (Hint: where do builders place the reinforcing steel bars (rebar) when they make concrete pillar or posts?

Usually the students think of providing protection, like bumpers, in case something bumps into the post, but some students may realize it helps to provide stiffness.

→ The arrangement of these cells demonstrates a mechanical principle evolved in plants (and used by engineers). Placing reinforcing steel bars (rebar) in the corners and/or around the periphery of a column provides the most strength using the least amount of material. This also explains the pattern of vascular bundles in a ring. Xylem provides support as well as transport for water and mineral nutrients.

Figure 2.12
A. Magnified cross-section of monocot root.
B. Closeup.

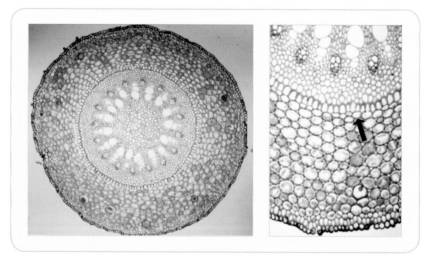

Inquiring About Plants

Roots also have characteristic patterns. This is a typical monocot root. Make a sketch of the general pattern.

A typical sketch will look like Figure 2.13.

•▸ What should we label the outer layer of cells?

Epidermis.

•▸ The outer layer is the epidermis and there is evidence of root hairs, particularly on the upper left.

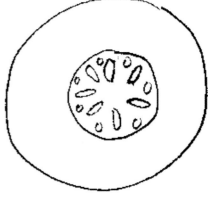

Some students may also notice that there is another distinctive single layer of cells just under the epidermis. This layer is easier to see on the close up. It is called the hypodermis and in this case it has "U-shaped" (actually cup-shaped) wall thickenings that are oriented inversely to those more visible in the endodermis. Hypodermis is not always present in roots, but it can be quite conspicuous.

Figure 2.13 Typical student sketch of magnified cross-section of monocot root.

•▸ Describe the thick layer of cells beneath the epidermis and hypodermis. Do they look like any cells we saw in the stem?

These are big cells with thin walls and large vacuoles that have intercellular spaces between them. They look like ground tissue in the monocot stem.

•▸ This layer of ground tissue, between the epidermis and central vascular tissue is called cortex, just like in a dicot stem. Now describe the cells that form the conspicuous ring just inside of the cortex.

These cells have thick "U-shaped" walls that stain a different color from the other cells.

•▸ This inner ring of cells is called endodermis. In monocots each endodermal cell has a distinctive "U-shaped" (actually cup-shaped in 3-D) thickening that contains a fatty material. If these cell walls are filled with a fatty material, how will this affect diffusion of water and dissolved substances from the cortex into the vascular tissues at the center where they can be transported to the rest of the plant?

The endodermis must block movement of water and dissolved substances.

•▸ Look carefully at the close-up of the endodermis where the arrow is pointing. Is there anything different about the two endodermal cells above the arrow?

The thickened fatty layer is missing.

↠ In monocots, occasional endodermis cells lack the special thickening on their inner walls. These are called passage cells, and they control what can move across the endodermal layer.

Inside the endodermis are two rings, one containing files of large cells and the other composed of bundles of smaller cells. Describe the appearance of the larger cells. Is their arrangement similar to anything you've seen in stems?

These cells appear empty and have a smooth red-staining wall. Their diameter is larger toward the center and smaller toward the endodermis. They are similar to the files of xylem in the vascular bundle of a dicot stem.

↠ These files of cells are bundles of xylem. What to you think the smaller green bundles of cells might be that alternate with the files of xylem?

The smaller bundles look like phloem cells.

↠ In roots, xylem and phloem are in separate bundles and bundles alternate in a ring: xylem, phloem, xylem, phloem, etc. Describe the appearance of the cells in the very center of this root.

These cells are fairly large, with thin walls.

↠ The very center of the monocot root contains ground tissue similar to the dicot stem. This is the pith.

The image below shows only the center of the dicot root. The epidermis and cortex are essentially the same as in the monocot. Make a quick sketch of the general tissue regions.

Figure 2.14
A. Magnified view of dicot root vascular tissue. B. Typical student sketch.

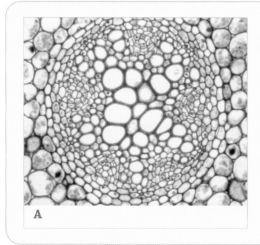

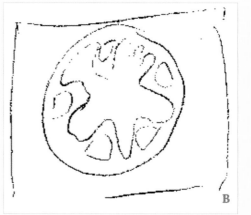

◈▸ Describe the cells at the very center of this root. What kind of cell do you think they might be?

> *These large empty looking cells have smooth, thick cells walls. The largest cells are near the center and smaller cells radiate out in a star-shaped pattern.*

◈▸ In dicots, xylem forms a conspicuous star-shaped pattern in the center of the root with the largest cells in the center and the smaller xylem cells at the tip of each arm. The smaller xylem cells began as separate bundles but the older xylem later joined in the center.

   Small patches of cells alternate with the arms of xylem. What do you think they are?

> *These look like enlarged phloem bundles from the monocot root that alternated with the xylem.*

◈▸ Between the arms of xylem are conspicuous round patches of green staining cells, these are phloem bundles. The phloem and xylem arms are separate and alternate as in the monocot. This is a characteristic of roots.

   There is a distinct layer of cells separating the xylem and phloem at the center of the root from the cortex cells around the outside. Based on the monocot root, what must this layer be?

> *Endodermis.*

◈▸ The endodermis is a distinct layer of smaller cells surrounded by larger cortex cells with large central storage vacuoles. There is not a thick "U-shaped" fatty thickening, only a thin fatty strip that wraps like a rubber band around the radial cell walls. On the right is an enlargement of one of the endodermal cells with the fatty strip clearly visible. The red-stained strip is very in the cross walls, and just barely visible running through the top or bottom wall that was included in this section.

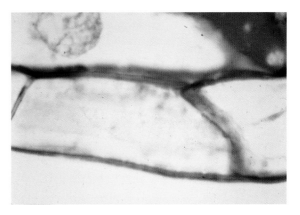

Figure 2.15
Magnified view of
dicot endodermis.

◈▸ What is the significant of the location of this hydrophobic material in the endo-dermal cells of the roots?

   What is the advantage of having the xylem in the middle of the root and not near the edge of the root, which is actually much nearer to the water and minerals in the soil that will enter the xylem?

   How does the form (large, open cells) of the xylem help its function?

   What is the significance of having dead xylem and living phloem?

This section focused on recognizing the patterns of different cell and tissue types characteristic of roots and stems, and associating structural pattern with physiological function. Now for a big challenge.

Consider both the root and stem patterns of either a monocot or a dicot. What is a major problem a whole plant must face given these patterns? Hint: what must happen so water can flow from the xylem of a root through the xylem of the stem?

Testing Hypotheses of Evolutionary Relationships of Ferns

In a nutshell: Even though your students may have little or no knowledge of the evolutionary relationship of one species of fern to another, they can make some predictions based just on general appearances of the plants. This activity begins with some general background on morphological differences among fern species. It then provides a paper-and-pencil example of the kind of algorithms used with the aid of computers to construct phylogenetic trees (hypotheses showing evolutionary relationships among species). Using a variety of living ferns in the classroom is preferred, so student groups could analyze a species in detail. Information is included here on four species that could be substituted for the real thing. Eventually, whether teams score morphological characters of live plants or work with coded data, students reconstruct phylogenies based on morphological, molecular, and combined data sets. This repetition helps students build "tree-thinking" skills. But for now, open with a general question: How do we test evolutionary relationships among species?

In the following, statements and questions for students are in roman and background information in sans serif.

▸▸ Four species of ferns, A-D, are shown in the photo (Figure 2.16). Take five minutes to discuss with your group how these ferns might be related, then construct a diagram (like a family tree) to illustrate which ferns are most closely related to each other and which are more distantly related. Be able to justify each of the links you make between species.

> Each group has just produced a hypothetical phylogenetic tree of the four fern species. Now they can test their tree more rigorously as the class constructs a consensus tree using several sets of data.

▸▸ Without knowing it, you have each constructed a hypothesis to show the relationships among the four ferns. What we now want to do is more rigorously test your hypotheses using real data and scientific reasoning. Scientists use a variety

Figure 2.16 Four species of ferns. A. *Psilotum*. B. *Lygodium*. C. *Osmunda*. D. *Pteridium*.

of different kinds of evidence to test evolutionary hypotheses, but morphological and molecular data are most common.

We will first learn how to collect such data, and then learn how to use the data to construct likely relationships. The first step is to describe potentially useful characters.

Morphology is the traditional basis for phylogenetic classifications and it continues to provide strong evidence for evolution. The diagram, table, and figures below illustrate some of the many morphological characters that are used to study fern evolution. First, we will go through the characters together.

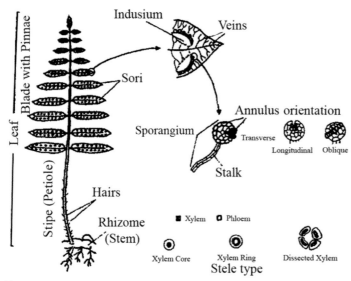

Figure 2.17
Diagram of some morphological characters to study ferns.

If you have access to living ferns, student groups could be tasked with examining one species to score the morphological traits for that species. Students may notice interesting (and potentially informative) morphological characters in addition to those described below. Some characters are easily visible to the naked eye, and others require dissection of plant parts and/or magnification. Many characters are typically included when constructing hypotheses of evolutionary relationships. For instance, Pryer and colleagues (2001) scored 136 morphological traits for ferns.

Table 2.1 Seven morphological characters of ferns

Character	Description of possible character states
1. Leaf Development	Developing leaves spiraled or not
2. Leaf Blade Shape	Simple or pinnate
3. Leaf Venation	Simple or branched (reticulate)
4. Fertile Fronds	Similar to sterile fronds or different from sterile fronds
5. Indusium	Present or not
6. Sporangial Dehiscence	Longitudinal or transverse
7. Arrangement of Xylem and Phloem (stele)	Central core vs ring or dissected

If you don't have access to living ferns, students who are comfortable working with primary literature could be tasked with finding the information from the Supplementary Materials from Pryer and colleagues (2001). To allow your students to work through this pencil-and-paper activity without access to live plants or primary literature, you can also provide the morphological character descriptions for each species after a class discussion of key morphological characters of ferns. The characters and species used here were selected to: 1) emphasize fern morphological diversity, 2) demonstrate how morphological characters can be coded for use in phylogenetic analyses, and 3) highlight the phylogenetic utility (or not!) of shared derived character states, character states found in only one taxon, and character states found in all sampled species.

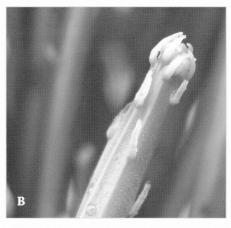

Figure 2.18 Developing leaves spiraled or not.

Figure 2.19 Leaf branching patterns, clockwise from upper right: simple, pinnate, multipinnate, and twice pinnate.

Figure 2.20 Leaf veins branched or unbranched.

⇢ As new leaves develop, they may be tightly spiraled or flat (Figure 2.18).

The overall shape of the blade varies from simple to more and more complex levels of branching. In ferns, leaflets are called pinnae. So, the levels of branching are described by levels of pinnae formed: pinnate, bipinnate, tripinnate, multipinnate (Figure 2.19).

Veins also have a distinctive appearance in the blade. In the simplest case the veins are unbranched, but veins that branch once or multiple times are most common (Figure 2.20).

When in a reproductive state, fertile fronds may be produced that are distinctly different from the sterile fronds, either sent up on different leaves or the frond divided into fertile and sterile regions. Although this is not especially common, it is easily seen (Figure 2.21).

Sporangia may or may not be grouped into sori. These sori may be covered by

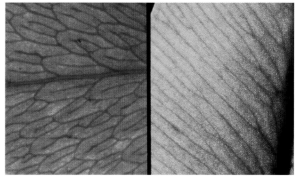

the leaf margin or other flap of tissue (an indusium) or not, on the undersurface of the leaf (Figure 2.22). And when a fern sporangium is ready to disperse its spores, it may dry and split open along various orientations (see Figure 2.17). In our sample of four ferns, dehiscence is longitudinal or transverse.

Figure 2.21 Separate fertile and sterile fronds (A) or not (B).

Figure 2.22 Covered or naked sori.

Stele is a term used for the arrangement of the vascular tissues, particularly in the stem. Xylem (usually red-stained in slides) and phloem (usually blue stained) form characteristic patterns in prepared microscope slides.

Inquiring About Plants

Now that you are familiar with some morphological characters for the four fern species, you might have some ideas about how they can be used to test evolutionary relationships. But first, let's introduce how to code characters.

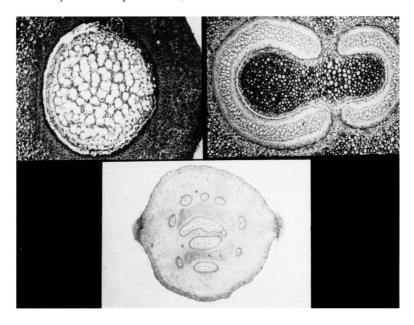

Figure 2.23 Solid, ring, and dissected vascular bundles (stele) in the rhizome or stem.

CONSTRUCTING A NUMERICAL DATA MATRIX

The ancestral condition of a trait, the condition present in the outgroup (in our case, *Psilotum*), is assigned the value 0. Any changes from the ancestral condition are assigned sequential whole numbers (1, 2, 3, etc.). For example, *Psilotum* has flat, non-spiraled, leaf development. The other taxa in our set have leaves that develop in spirals and they are coded 1 (Table 2.2).

Table 2.2 **Example of class data of morphological character state matrix.**

Character	Psilotum	Lygodium	Osmunda	Pteridium
1. Leaf Development	0	1	1	1
2. Leaf Blade Shape	0	1	1	1
3. Leaf Venation	0	1	1	1
4. Fertile Fronds	0	0	1	0
5. Indusium	0	1	0	1
6. Sporangial Dehiscence	0	1	0	1
7. Arrangement of Xylem and Phloem (stele)	0	0	1	1

If you don't have access to live plants, here is the completed data matrix your students can use to reconstruct phylogenetic relationships among our ferns. Otherwise, your student can create and complete their own data matrices for species they examined. Then the data for characters states of all species can be combined into a class data matrix, which should look something like Table 2.2.

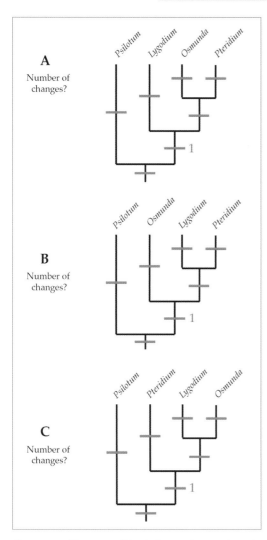

A
Number of changes?

B
Number of changes?

C
Number of changes?

Figure 2.24 Three possible phylogenetic trees based on morphological data. One character change is mapped on the trees: the change from non-spiraled leaves to spiraled leaf development on the branch where *Lygodium, Osmunda,* and *Pteridium* last shared a common ancestor with *Psilotum*.

CONSTRUCTING A PHYLOGENETIC TREE

➡️ Like any good tree, a phylogenetic tree must have a root, the origin from which the branches of the tree grew. The outgroup is used to root your phylogenetic tree. Taxa are added sequentially to the root to provide the most likely tree—based on the data you have used to construct the tree.

Let's start by mapping using the morphological matrix you generated above to find the tree with the fewest number of changes. From our matrix we know that *Lygodium, Osmunda,* and *Pteridium* all have spiraled leaf development, whereas *Psilotum* does not. We can therefore assume the change from non-spiraled to spiraled leaf development occurred on the branch where *Lygodium, Osmunda,* and *Pteridium* last shared a common ancestor (see 1 in trees A, B, and C in Figure 2.24). Now map the distribution of changes from the other seven characters onto the trees A, B, and C. When you have finished, count up the total number of changes for each of the trees. The preferred tree—the most parsimonious tree—is the pattern of phylogenetic relationships requiring the fewest number of changes.

Which of the three trees is the most parsimonious for the morphological dataset?

It should be tree B with a total of 8 changes. In addition to leaf development (character 1 from Table 2.2), students should indicate that leaf blade shape and leaf venation changed on the branch where *Lygodium, Osmunda,* and *Pteridium* last shared a common ancestor with *Psilotum* (i.e., note 1, 2, and 3 beside that branch). *Lygodium* and *Pteridium* share indusium and sporangial dehiscence characters in common, so students should note characters 5 and

Inquiring About Plants

6 beside that branch. Character 7 is present twice, on both the *Pteridium* and *Osmunda* branches. And character 4 occurs only on the *Osmunda* branch. Trees A (9 changes) and C (10 changes) have more changes. So, they are less parsimonious.

⟶ Next we will move on to molecular traits, and also use these data to make our phylogenetic trees. Molecular traits have become increasingly important in determining phylogenies.

> One of the important reasons for this trend is that the data are much less prone to interpretational bias by the investigator. The results are also easily quantifiable which again aids in eliminating bias from the interpretation. An added benefit is that whole specimens are unnecessary. Frequently only a small bit of tissue is required and in some cases even fossil material may be used to obtain molecular data!

⟶ Depending on the purpose of the study, it may be more useful to examine DNA from the nucleus, chloroplasts, or mitochondria. DNA sequencing using Polymerase Chain Reaction (PCR) is now highly automated allowing the collection of vast amounts of data in a very short amount of time. Other methods are being developed all the time to increase the speed and accuracy of DNA sequencing. The data output are strings of nucleotide sequences from specific genes or multiple genes across the genome.

> For students who are interested in taking the bioinformatic analysis a step further, sequences can be downloaded from the supplementary data from the Pryer (2001) paper or accessed by individual genes for particular taxa in GenBank.

⟶ Below is a table with DNA sequence data from our four fern species.
Compare the sequences for the four taxa, position by position. Nucleotides identical to those of *Psilotum* are considered to be ancestral and not changed. We will now use these data to reconstruct evolutionary relationships.

Table 2.3 DNA sequence data from our four fern species.

	1	2	3	4	5	6	7	8	9	10
Psilotum	A	C	C	C	G	T	A	G	A	G
Osmunda	A	G	C	A	C	G	C	G	A	A
Lygodium	C	G	G	A	C	T	C	G	C	A
Pteridium	C	G	C	C	G	G	A	G	C	A

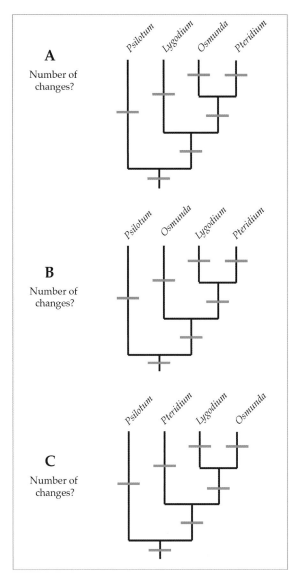

A

Number of changes?

B

Number of changes?

C

Number of changes?

Figure 2.25 Three possible phylogenetic trees based on molecular data in Table 2.3.

Now use the same principles to reconstruct the pattern of changes in your molecular data set to find the most parsimonious tree.

In this case, it should be tree C with 13 changes.

●▸ It is common for evolutionary biologists to use a combination of morphological and molecular datasets to estimate phylogenetic relationships using a "total evidence" approach. You can estimate the total evidence phylogeny by mapping the morphological and molecular characters onto the trees together, or simply adding the number of changes from your morphological and molecular reconstructions. Which of the three topologies is the most parsimonious total evidence estimate of relationships?

It should be tree B with a total of 8 morphological changes + 14 molecular changes = 22 total changes.

●▸ Fortunately, there are computer programs that perform the necessary calculations and comparisons to generate trees for large data sets. They have much more sophisticated "rules" than we used, but the basic process is the same.

A phylogeny of ferns by Pryer and colleagues (2004) is shown in Figure 2.26. Our outgroup (*Psilotum*) is in the whisk ferns (near the bottom of the figure). Ask your students to examine the figure to see if the pattern of relationships in this much larger analysis is consistent with the morphological, molecular, or total evidence phylogenies they have reconstructed.

Resource:
GenBank: http://www.ncbi.nlm.nih.gov/genbank/

Inquiring About Plants

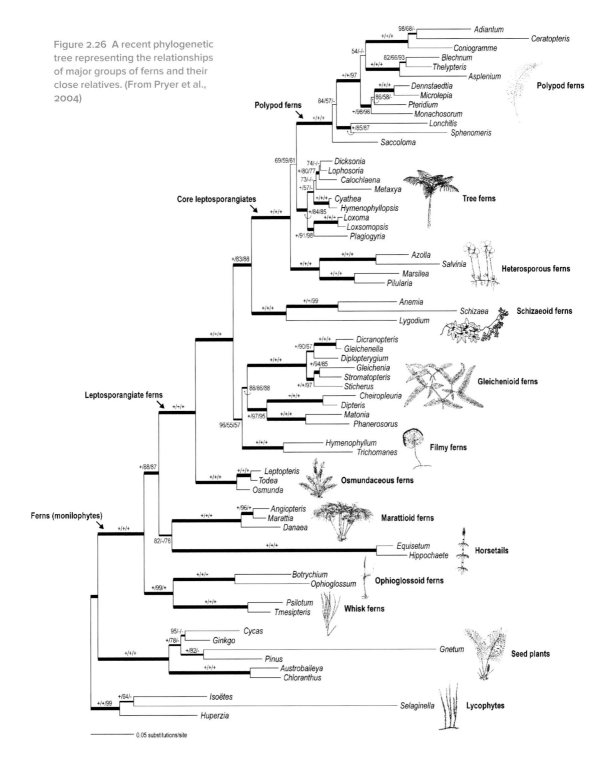

Figure 2.26 A recent phylogenetic tree representing the relationships of major groups of ferns and their close relatives. (From Pryer et al., 2004)

2.2. Focusing on Student Learning

The biggest change any teacher can make to improve his/her instruction is simple—ask yourself before every class, how are you going to help your students "learn" biology instead of what are you going to "teach" today? Please consider what this simple refocusing does—it converts a class from being teacher-centered to one that is student-centered, with many benefits. Most of us worry about what we have to "cover" in our courses—you still have to decide what topics and concepts to teach, and you can choose what you want your students to study. However, if you are worried that students simply memorize information, regurgitate it back to you on an exam, and then quickly forget what they memorized, then we might need a teaching intervention! So, how can you help your students "learn" and understand biology better? First, give them more time to reflect on what you "cover" and to actually work with the information they are given so they can gain a deeper understanding of the material. This "time on task"—the task trying to understand the information—will help your students learn whatever you choose to teach. You can "feed" your students any biological information you want, but you need to give them time to digest that information, otherwise they will just regurgitate that information on the exam, and soon forget!

There has been much research conducted over the last decade about how students learn, in general, and how students learn science, in particular. Two informative books on the subject are *How People Learn: Brain, Mind, Experience, and School* and *How Students Learn: Science in the Classroom*. Both of these are published by the National Academies Press (www.nap.edu).

We should all try to incorporate lessons we have learned about learning throughout our courses. This includes how we can help students recognize what they do and don't know and how they can help themselves learn and understand (metacognition) and to help them recognize their strengths and weaknesses related to specific concepts before they take a test that counts toward their grade (using formative assessments).

FORMATIVE ASSESSMENTS

If you want to check your student's understanding of a topic before an exam, consider using "concept inventory" questions. Concept inventories are banks of diagnostic questions that are intended to be used to assess a students' level of understanding and to reveal and remediate misconceptions they might have about particular concepts in biology. These questions are intended to be used in a formative way, without students receiving any "penalty" for answering questions incorrectly. The questions are also intended to generate meaningful discussions among your students. However, they are definitely NOT intended to be used as part of a summative evaluation (for a grade). You can learn more about concept inventories by reading D'Avanzo's (2008) "Biology Concept Inventories: Overview, Status, and Next Steps." Figure 2.27 is a question from that paper (a is correct answer):

Figure 2.28 is another question, this one is from Ebert-May and colleagues on "Disciplinary Research Strategies for Assessment of Learning."

Figure 2.27 Where did all the fat/mass go? A diagonstic question cluster to identify problematic patterns in student thinking about challenging biology content. (From D'Avanzo, 2008)

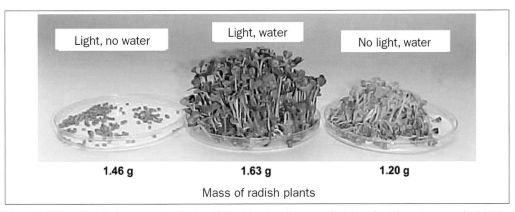

Figure 1. Radish problem that was presented in class: "In the laboratory, three equal batches of radish seeds are weighed at 1.5 grams (g) each. Two batches are watered and allowed to germinate (radish seeds germinate rapidly). One of these batches is put in the light and the other in the dark. The third batch is not watered and is left in the light. The dry biomass for each batch is measured after 10 days. Predict the biomass of the plants in each treatment after 10 days." In groups, students discussed the problem and made predictions. The instructor recorded predictions from a sample of randomly selected groups. Then the results of the experiment were shown to the class, and each student wrote an explanation.

Figure 2.28 Predicting the biomass of radishes in three treatments. (From Ebert-May, Batzli, and Lim, 2003)

Note: You don't give the students the data until AFTER they have discussed the problem and have predicted an outcome for all of the batches. Ask them to place the three trays of seeds/seedlings in the correct order based on their **dry** weight—from lightest to heaviest. Students rarely get the correct answer the first time out.

DEVELOPING YOUR OWN QUESTIONS FOR DISCUSSION

You can develop your own questions that can be used in the same formative way. Identify questions on your exams of the past that many of your students answered incorrectly. Can you develop distracters based on their incorrect answers and write a multiple choice question that you use for discussion?

Note the following two questions—the first is one you DON'T want to use to generate discussion, because it has only one "correct" answer. The second is better because there is room for debate and disagreement among students—and it gets at the misconception that plants get food from the soil.

Question 1	Question 2
Which of the following is the plant family for oak trees?	How much food does a plant get from the soil?
a) Fumariaceae	a) all of it
b) Rosaceae	b) some of it
c) Fagaceae	c) none of it
d) Rubiaceae	d) depends on the plant species

For such questions, ask students to choose an answer on their own. Then they should write a one-sentence reason why they chose the one particular answer. As a group, ask students which answer they chose, and then give them a few minutes to convince their peers that they are right and their fellow classmates are wrong. See if any students have changed their mind, or their explanation for their choice!

To help students learn, they must understand how to learn and what it is they already know. There are different ways you can tell if students truly understand a concept or if they have simply memorized information. The simplest way is to ask them to explain a concept or topic in their OWN WORDS—can you ask a question about the topic in a slightly different way than the way you taught the concept so that students can't just repeat what they have heard or read? Also, can they teach the concept/topic to someone who is unfamiliar with the subject matter—teaching a topic is one of the best ways to learn about it.

You should emphasize to students that if they try to make sense of information, then they don't have to memorize the material—they can figure out answers to questions because they understand the material. And, as we all know, our memories will fail us every time!

Here is a short activity to illustrate how important it is for students to look for meaning in the information they are trying to learn—instead of just memorizing it. Students too often look at scientific information and simply try to memorize words and definitions without having any ability to interpret or explain the information. The point of the following is this: if you make sense of the information you are studying, you don't have to memorize it—and you can retain a lot more information for a lot longer.

Tell your students that you are going to ask them to look at a set of words about Christopher Columbus, and that they are going to be given 20 seconds to try to remember as many of the words as they can. It is important to note that both groups are going to be looking at the SAME SET of words—just arranged differently.

Divide your class in half. One half of the students will be shown the first slide for 20 seconds. The other half has to keep their eyes closed during this time. Then, the first half closes their eyes, while the second half views the second slide. At the end of this second viewing period, all students are asked to write down as many of the words that they remember from the particular slide they viewed—give them a minute to write down as many words as possible. Tally up and average the number of words that each group recalls—there should be a significant difference in the number of words that the two groups remember. Get them to discuss what the significance of this is. Students who recognize a pattern in the words (the words are formed into a story that makes sense) can quickly "absorb" the information and have a much greater ability to recall that information compared to students who simply try to memorize the information without any understanding.

 # 1 Set of Words

Now	Sought
Three	Proof
Sturdy	Forging
Sisters	Along
Set	Through
Sail	Calm
And	Vastness

 # 2 Set of Words

Sisters	Set
Proof	Forging
Calm	Vastness
Sturdy	Three
And	Sail
Along	Sought
Now	Through

REVEALING AND DEALING WITH MISCONCEPTIONS

As noted in the section on Communication, drawings can be useful in helping students "synthesize their ideas" about a concept. Drawings can also help them understand what they don't know and let you know what misconceptions they have. Once you have identified misconceptions, then you can work to help students construct a more appropriate and accurate understanding of biological phenomena. And remember, as students try to construct their understanding (constructivism), if they possess misconceptions, these misunderstandings will form an unstable foundation on which they cannot build greater knowledge about a subject. One misconception that students have is how plants incorporate carbon into their bodies.

Start this activity by talking about humans as being "carbon-based life forms" as noted in Star Trek (although this reference may be lost on your students). Humans, and all other forms of life on Earth, are composed of biological molecules (fats, carbohydrates, proteins, and nucleic acids) that all contain carbon, hydrogen, and oxygen atoms—and almost all of these biological molecules started as the sugar that plants produced in photosynthesis. Make the claim that there has always been the same number of carbon atoms on Earth since life evolved on our planet. That means that long-dead plants and animals contained carbon atoms that could now be part of the bodies of your current students. So, divide your students into small groups, and get each group to illustrate, on a large piece of flip-chart paper, how the carbon atoms of a *Tyrranosaurus rex* dinosaur that lived 68,000,000 years ago got into their body. Students can use illustrations and a few words, but they must show the route that a carbon atom took.

Students will invariably draw a dead dinosaur that decomposes into the ground. They will then show that the plants took up the carbon from the ground and incorporated it into their bodies. Eventually, the carbon atom from the plant was eaten by some animal, and that animal was eaten by one of your students (understanding a long time has passed between the time the dinosaur died and the time the student had lunch). This drawing, however, indicates that most students forget about the important role that decomposers play in the cycling of nutrients in our world. Plants don't take up carbon from the soil—they use the carbon dioxide in the air as their source of carbon for producing sugar during photosynthesis. So, carbon from the dead dinosaur in the ground must somehow get into the air before it is useful to plants. Have students display their posters, and let students critique each team's drawing.

Now, as an example of giving students more time to digest this information, get students to provide a biological meaning to the biblical quote, "all flesh is grass," identifying how their biological molecules first began in a plant and ended up in their bodies. Finally, you might get them to consider this question. . . . "Have you ever been through stomata?"—if all of their biological molecules started as sugar made by plants, then the answer is, "Yes!"

Is an Apple (or any other fresh fruit or vegetable) Alive?

In a nutshell: Surveys suggest that the majority of Americans do not think that plants are alive, and certainly not parts of plants that have been picked and stored for any period of time. This activity will challenge the misconception that fresh fruits and vegetables (and plants in general) are not alive.

In the following text, suggested teacher script is in roman noted with a bullet, procedures indicated in roman, and alternatives for elaboration in sans serif.

Bring an apple to class. Just before starting class, take a conspicuous bite of the apple, chew it and swallow. Now ask the class to write down the name of a living organism, then quickly survey the class to generate a list of 10–20 of the organisms they listed. The chances are than more than 90% of the class will list a vertebrate animal of some kind. Now ask them:

►► Do you think the apple I'm eating is alive or not?

Do a quick tally of the class response. Typically, the majority will vote "no," but regardless of the student vote, the next step is to challenge the class to describe how to recognize life.

◆▸ We've just done an opinion survey. Now, how can we test which opinion is most likely correct? What are some criteria of living things that we can recognize as being distinctive from non-living objects?

Suggest that the class make a list of characteristics of living things that you can put on the board. Your task is two-fold. First will be to prompt the class with questions that will lead to a list of criteria. However, just as important will be to provide examples of EXCEPTIONS to each of the criteria they list.

Some typical criteria might include:

Living things breathe.   This is a typical animal bias that you will want to try to work around to the concept of **cellular respiration**.

Living things move.   This is another animal bias that usually refers to locomotion of the entire organism. Some notable exceptions to mention are sessile adult stages of many invertebrate animals, such as corals, and motile reproductive stages of many plants (algae and non-seed plants). Scale is also important. Time-lapse photography allows us to visualize many plant movements that are too slow for us to recognize 'by eye."

Living things reproduce.   Distinguish between cell reproduction and organismal reproduction; sexual vs. asexual reproduction. Students will normally be thinking only of sexual reproduction in animals.

Living things are composed of cells.   Viruses are not cellular, and biologists are split about whether or not to consider them to be alive.

Living things grow.   Growth can be divided into a number of components, both at the cellular and organismal levels, including: cell division; cell enlargement; and cell differentiation.

Living things respond to stimuli.   Ask for some examples—most will be animals.
Protists can be chemo-, photo-, and thigmotropic (response to touch) and slime molds also respond to temperature and humidity shock.
Fungi can be phototropic, gravitropic, and hydrotropic.
Plants are phototropic, gravitropic, thigmotropic and respond to light/dark cycles.

Living things obtain food.   Fungi and carnivorous plants secrete enzymes to externally digest food compounds, then absorb the nutrients. Plants are autotrophic and produce their own food.

For additional ideas see Wilson and Mullins (1980).

With a list of characteristics of living things, you're now ready to start designing experiments to test which, if any, of these criteria apply to your apple.

⇢ As you can see by our list of some characteristics of living things, there are always exceptions. This is why there is no single definition of life. Instead, we must evaluate if all or most of the characteristics are supported.

> If you are familiar with concept mapping, this provides a convenient tool for identifying possible tests. Each of the criteria listed are concepts included on the map. The connecting lines (propositions) suggest the relationship between concepts. Virtually every proposition is a hypothesis that could be tested with an experiment. Divide the class into small groups and have each group choose three possible criteria to test.

If you don't use concept maps, divide the class into small groups and ask each group to choose three of the criteria listed that could be applied to your apple.

⇢ Every group has chosen three criteria to test; now we'll decide which criterion your group will be responsible for. You will have to formulate a hypothesis and then design an experiment to test if the apple meets your criterion for life—do you support or reject your hypothesis?

Certain criteria will be more popular than others so by having each group consider at least 3 alternatives, you should be able to get a variety of different criteria examined with minimal duplication.

Some examples are:

Cellular structure is a characteristic of living things. Therefore, if the apple is alive it should be composed of cells.

> The appropriate experiment will be microscopic examination of thin, stained, hand sections of a piece of apple.

Cellular respiration provides energy for living cells. Therefore, if the apple is alive, it should undergo respiration.

> The appropriate experiment will be to test for respiration either by determining O_2 consumption or CO_2 production.

Living cells respond to stimuli therefore; if apple cells are alive, they will respond to stimulus.

The cytoplasm, and cell membrane, will respond to different osmotic solutions if cells are alive. Plasmolysis will occur if the cells are placed in salt water, and the cells will rehydrate if salt water is replaced by distilled water. Microscopy will be necessary to observe these responses.

Living things move would include cytoplasmic streaming within cells. Therefore, if apple cells are alive, streaming should be evident in the cells.

Microscopy will be necessary to observe these movements, if present.

Living things reproduce, therefore, if the apple is alive, the cells may be able to divide and reproduce.

The appropriate experiment will involve placing cells in tissue culture, then looking for evidence of cell division—either directly (mitosis and cytokinesis), or indirectly by cell counts.

After students have had some time to design their experiments, ask for a volunteer group to describe their plan. Cellular structure is most straight-forward, so ask that group to volunteer. The main thing to look for is whether or not they included a control. Without prompting, students almost NEVER include a control for any of these experiments.

➼ What would be an appropriate control for any of your experiments? If you don't know for sure that your apple tissue is alive, is there any way you could know for sure that it is dead?

Some alternatives students might suggest for ensuring tissue is dead include:
- Boiling (for a few minutes)
- Pressure cooking or autoclaving (15 min. is "standard" time at temperature/pressure)
- Microwaving (for a few minutes)
- Soaking in 70% alcohol (for a few minutes)
- Soaking in 10% household bleach (for a few minutes)
- Freezing (overnight)

For each of the above methods a time is suggested, but you can leave this open for a group decision.

Following are some suggestions for typical student experimental designs.

Remember, they will want to do the same treatment with their "killed" tissue and the "fresh" tissue. Microscopy will be necessary for several of the experiments.

Cellular composition is the most straight forward and this group will probably be the first to have results. **Hint:** If students have results well ahead of the rest of the class, challenge them to describe these cells as fully as possible—including an estimate of how large the cells are and a careful examination for visible structures.

If heat was used for killing the control, there will be visible differences between the control and fresh tissue, however, if chemicals were used, there may be no obvious differences and the results will be ambiguous.

Response to stimuli (different osmotic concentrations) is also straight forward. This group will be able to get useful results regardless of how the control tissue was killed. The first step will be to demonstrate plasmolysis, which requires the cell membrane to be functional (living cells). Don't let them forget to do the confirmation experiment of deplasmolysis using distilled water. **Hint:** plasmolysis will be easiest to observe if the tissue has colored cells—either pigments in the vacuole (epidermis of many fruits), or pigmented organelles (chloroplasts in green fruits and vegetables or chromoplasts in yellow or orange tissues). **Hint:** epidermal peels of most fruits and vegetables will have visible stomata, which will respond readily to different osmotic concentrations.

All living cells will respond, no killed cells will, regardless of killing treatment.

Movement in cells. Guard cells, see above, will move in response to stimulus— light/dark in addition to osmotic concentration. Students can also look for streaming in the cytoplasm of cells.

It is unlikely that students will observe any streaming, although it is there. With thin sections and a good microscope (400X or higher) set for optimal resolution (focused condenser and iris diaphragm closed down), you can just barely see mitochondria as small black "dots" around the periphery of most living plant cells. In most cases, however, the results will be ambiguous.

Respiration. Depending on the technique chosen for showing respiration, this will be relatively quick and straightforward. Using Phenol Red is an easy, fast, and cheap way to demonstrate respiration. An Oxygen and/or CO_2 meter is also quick and easy and provides quantitative data. Respirometers are also quantitative but require more set up and care than a direct measuring meter.

Students will get results, fairly quickly, that differentiate between the tissue sample and the killed tissue. If they are finished far ahead of other groups, give them the following challenge question.

•▸ What alternative explanation would give you exactly the same results? Hint: Why should you wash?

> An alternative explanation for evidence of respiration in these samples is that bacteria on the surface of the apple pieces are respiring. Any control treatment would also have killed any surface bacteria, so the absence of respiration here also is explainable.

•▸ How can you redesign your experiment to eliminate possible effects due to surface contamination?

Most students will come up with washing, perhaps with an antibacterial soap, and/or surface sterilization with alcohol or bleach. Either of these must be for a shorter time than to kill any potentially living cells in the tissue sample. Usually we follow surface sterilization with several rinses of sterile distilled water (autoclave, pressure cook, or at least boil a half-filled flask of distilled water with the mouth covered with aluminum foil, then let it cool).

> **Growth or cell division.** Any groups choosing to test this will have to do a long-term experiment: tissue culture. Sterile technique is time consuming and requires meticulous attention and the results will not be apparent for several weeks. If you have a particularly good team, this would be a challenging long-term project for them.

END OF LAB SUMMARIES

Several of the groups will be able to share results by the end of one or two lab periods. **Cell structure** and **respiration** are the most likely to have results unambiguously supporting that the apple tissue is alive. The results for **response** and **movement** will vary considerably depending on what material is examined and what characteristics the students choose to examine.

It is UNLIKELY that all of the tests will support the hypothesis that the tissue is alive—however, the majority will. This is a good learning point for students. When scientists do experiments, their results are often ambiguous and require that the experimental procedures must be modified or even that alternative approaches must be tried.

Virtually all of the fresh produce we find in the store is alive—even if it was picked weeks or even months ago. Consequently, you are eating living flesh whenever you munch on fresh fruit or vegetables!

Carbon Cycle, Photosynthesis, and Cellular Respiration

In a nutshell: This activity works best as a laboratory exercise because what seems pretty straight forward often produces results that are typically explained as "experimental error" and ignored. However, the results are real and provide you an opportunity to force students to look for, and test, alternative explanations for their data. Here it is presented as an in-class activity to promote deeper understanding of the carbon cycle in the balance of nature. Students should work in small groups.

➡ What would you need to set up a long-term freshwater aquarium that is not dependent on electricity in any way, e.g., no pumps or lights? For instance, if we already had an aquarium in the lab, how would you have to change it so we could remove the pump and lights and yet the aquarium is self-sustaining?

You'd have to set it near a window where there would be light and have both plants and animals in the aquarium.

➡ Make a list of all of required materials, both living and non-living you would need to do this. It might also be useful to make a sketch or diagram of what the set-up will look like. For each item listed, write out a brief explanation or description of the specifications for the material or its role in the completed aquarium.

Figure 2.29 Freshwater fish in a table-top aquarium.

Go around the room to ask individual groups to add to a class list of the materials that would be necessary, as well as an explanation for why it is needed.

These typically include: an aquarium, water, substrate of some kind, aquatic plants, and fish. Typically, the plants are necessary to photosynthesize and produce food and oxygen for the system and the animals cycle carbon and oxygen through cellular respiration.

➥ You now have the components necessary to begin a discussion of the carbon cycle, and more specifically the roles of cellular respiration and photosynthesis as driving forces in the carbon cycle.

Create a flow chart or diagram to represent the roles of photosynthesis and respiration in the carbon cycle within your aquarium. It might look like this:

Figure 2.30
Diagram of carbon cycle within aquarium containing plants and fish.

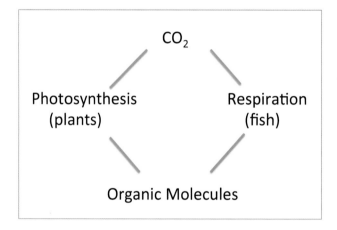

➥ Our diagram of the carbon cycle is a testable model of a mini-biosphere. How could you test to see if you have the right number of plants and animals to keep your aquarium sustainable?

Measure how much photosynthesis the plants do and have enough plants to equal the amount of respiration done by fish.

➥ In aquatic systems CO_2 dissolves in water to form carbonic acid, thus lowering the pH of the water. We can use pH to indirectly measure change in CO_2 related to the carbon cycle.

Design a set of experiments to measure the amount of photosynthesis done by a plant and the amount of respiration done by a fish.

Students will usually want to grow the plant with and without light to demonstrate photosynthesis. They may or may not think of testing the animal under the same conditions. They will probably NOT include the control of aquarium water alone without either the plant or fish. They will probably NOT include replications of any treatment.

Here is a sample experiment, with ten replications per treatment, done using fish and plants from the same aquarium. Each treatment was run for 50 minutes before pHs were recorded (assume that the initial pH of water from the aquarium was 7.0). The experimental design and results are presented below. Use this as a model to have your students set up their own experiment and gather their own data. Depending on class size and materials you could assign a specific treatment to each group and pool data, or groups could run the entire experiment and each group would be a replicate.

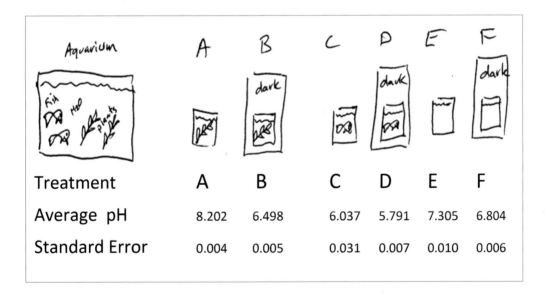

Treatment	A	B	C	D	E	F
Average pH	8.202	6.498	6.037	5.791	7.305	6.804
Standard Error	0.004	0.005	0.031	0.007	0.010	0.006

●▸ How do you explain the results of treatment A? Is this what you predicted? Why or why not?

Figure 2.31 Experimental design and results with replicates for light and dark treatments.

The pH went up so there is less CO2 dissolved in the water. This is because photosynthesis converted some of the CO2 into organic molecules. Yes, this is what we predicted.

●▸ How do you explain the results of treatment B? Is this what you predicted?

The pH went down a little bit. Most students will probably suggest that this was only a small amount and probably due to experimental error. Some students may realize that plants also undergo cellular respiration (both in light and dark), so more CO_2 would have been produced by the plants and the pH would go down.

⟢ How do you explain the results of treatment C? Is this what you predicted?

The pH went down quite a bit and this is what would be expected from respiration. More CO_2 is produced, which will lower pH.

⟢ How do you explain the results of treatment D? Is this what you predicted?

This has the lowest pH of all—even a bit more than C. Most students will argue that the difference is just experimental error.

⟢ How do you explain the results of treatments E and F? Is this what you predicted?

Students may argue that this demonstrates the amount of experimental error that could have been involved in B and D.

It is possible that experimental error is responsible for the small changes in pH found in treatments B, D, E, and F. However, the standard errors are very low in each case. If it was just random error between each of the 10 repetitions for each treatment, the averages should even out and show no change and the standard error should be larger.

⟢ Assume that the changes observed in E and F are real. What would have to be happening in each sample?

The pH in E has gone up slightly. The only other samples to increase were in A where photosynthesis was occurring. Thus, somehow photosynthesis might be happening. The pH in F went down slightly. In all of the other treatments this occurred because of respiration.

⟢ If the pH in E went up, this suggests that photosynthesis was occurring here, just like in treatment A. Similarly, if pH in F went down, this would be like all of the treatments where respiration occurred. What kind of organisms might be in the water that are capable of undergoing photosynthesis?

Bacteria and phytoplankton

⟢ What kind of organisms might be in the water that are capable of undergoing respiration?

Bacteria and both zooplankton and phytoplankton.

↠ How could we test if there are bacteria and/or plankton in the water that are affecting the results in this experiment?

The first suggestion is usually to use distilled water for the treatments. This can lead to a review of osmosis and especially animal cells. Another common suggestion is to use sterile water—and this is on the right track. What we want to do is sterilize aquarium water. This occasionally leads to students suggesting that we should also sterilize the containers and the macro-organisms we want to use. Great ideas, but this is difficult to do in practice.

3 Connecting Concepts And Investigating The Plant World

So far, you have seen a wide variety of activities for the classroom, lab, or outdoors that illustrate ways to integrate science process skills and concepts. You are now also armed with some of the literature on the effectiveness of inquiry and research on student learning. With this foundation in place, the stage is set to dive into the realm of deep understanding. What do you want your students to be able do with the knowledge and skills they have been building? What kinds of evidence do you look for that shows your students understand the course material—not simply remember it long enough to pass a test? The ability to connect and apply scientific ideas is strong evidence of understanding. You can be intentional in crafting opportunities for your students to practice these higher-level thinking skills and demonstrate their proficiency by translating their knowledge and skills to a new context.

There are several ways to select and sequence content for your course to help students connect biological information. Revisiting ideas throughout a course is one very powerful approach. Evolution is an excellent theme to return to repeatedly. As Dobzhansky (1973) eloquently said, "*Nothing in biology makes sense, except in light of evolution.*" Another approach for connecting concepts is to design lessons in which students work with information from more than one discipline. Any of the cross-cutting ideas in the Next Generation Science Standards (e.g., "structure and function" or "stability and change") are also good candidates for making connections across concepts, whether presented throughout a course or in an individual lesson.

The activities presented in the book give your students multiple opportunities to practice applying their knowledge and skills. Your students may not realize it, but this repeated practice is preparing them to be independent investigators. We close the book with ideas and supporting information you can use as capstone or culminating tasks for students to design and carry out their own inquiries about plants.

3.1. Connecting the Big Ideas

Themes are useful for organizing the facts and concepts that you teach. A biological theme can help students make sense of what is seemingly a lot of disparate information. The following are some examples of how to incorporate evolution as a theme in all parts of your course. If you do so, you will accomplish several things: first you provide a framework for understanding biology by connecting different facts and concepts. You also avoid simply teaching evolution as a single concept on a single day or two in your class. Third, by teaching about evolution early and often, you don't allow students to skip school or avoid learning about evolution if you teach it in just a few lessons. Mostly, however, when you use evolution as a theme, you help students make sense of biological information.

WHAT DO THE FOLLOWING ALL HAVE IN COMMON?

- Each year, we have to get a new flu shot.
- Prescriptions say that you must take all the antibiotics and not stop when you feel better.
- The Bush government spent over $4 billion dollars to prepare for a bird flu epidemic.
- It's hard to lose weight, and lower back problems are very common in humans.
- Many humans need teeth removed and braces to correct tooth alignment.
- Every organism on Earth has DNA.
- Why do we use rats and mice instead of snakes to study human medicines and disease?
- No fossils of modern humans are found in rocks older than 200,000 years. No fossils of dinosaurs are found in rocks younger than 60 million years.
- There are new superbugs causing diseases that cannot be controlled with currently available antibiotics.
- Ranchers and farmers are finding it more difficult to control weeds and pests with the same herbicides or pesticides that they have been using for years.

The answer is: Evolution!

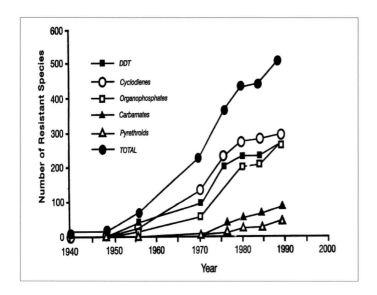

Figure 3.1 Increases in the number of pest species resistant to the principal classes of insecticides. (From Metcalf and Luckmann, 1994)

WHAT IS EVOLUTION?

Evolution is "a change in the gene frequency of a population over time," or "descent, with modification, of organisms from common ancestors." Evolution may happen very quickly (one generation), and evolution does not have to result in a new species. You might begin your discussion of evolution with artificial selection. For centuries, humans have been selecting which organisms can reproduce to develop more desirable fruits, vegetables, livestock, and pets. An example of artificial selection is the development of all of the following foods— from the same ancestor. Humans artificially selected for and bred wild mustard plants with certain attributes to develop such crops as Brussels sprouts, cabbage, cauliflower, and broccoli—all produced from the same wild ancestor, which then became part of the human diet. This is an example of descent, with modification, of organisms from a common ancestor—evolution driven by artificial selection because humans decided which organisms could reproduce. If someone had never seen these plants before, they would probably tell you all the plants were from different species because they look so different from each other. Actually, they all belong to the same species and are very similar genetically—which shows you how dramatic changes can be with a just a few differences in genetic makeup.

Inquiring About Plants

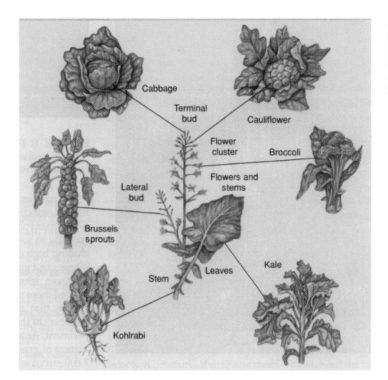

Figure 3.2 Common garden vegetables developed from a wild *Brassica* by artificial selection to enhance edible features of leaves, stems, and flowers. (From Uno, Storey, and Moore, 2001)

NATURAL SELECTION

Natural selection is differential reproduction—meaning that when you look at a population of organisms, not all of them will produce the same number of viable offspring (or contribute to their production equally). Some organisms in every natural population will produce more offspring than others. But it is "nature" and the environmental factors in nature that determine which organisms will reproduce and pass their genes on to the next generation of organisms. In general, natural selection is one of the main driving forces of evolution (Note: There is a misconception that natural selection and evolution are the same thing—they are not).

Every natural population possesses genetic variation (unnatural populations such as a field of wheat possess plants that are all genetically similar to each other)—natural populations are comprised of individuals with different genotypes, and different genetic makeup. What is the significance of this biological fact? Sexual reproduction leads to offspring that are genetically different from each other, and almost all populations of organisms use sexual reproduction to produce offspring (some plant populations reproduce some/mostly through asexual reproduction too). What's the advantage of reproducing sexually? For those organisms that produce offspring that are genetically different, at least some of their offspring will survive and reproduce even when the environment changes....and the environment will always change! These changes can be over a short period of time or a long period of time (such as global warming). Remember, however, when there is genetic variation (and phenotypic variation) in a population, the organisms that are favored in one environment may be at a disadvantage when that environment changes (as seen below).

Figure 3.3 The environment influences which organisms are favored. In this example, tall plants initially have a selective advantage over short plants, which are shaded. However, the addition of an herbivore that eats tall plants and leaves the short ones changes the environment and selective advantage. (From Uno, Storey, and Moore, 2001)

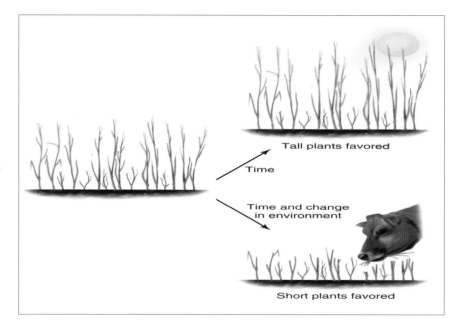

THE PRINCIPLES OF NATURAL SELECTION

Charles Darwin's four main ideas about natural selection are:

1. Organisms produce more offspring than can survive
2. This creates competition for limited resources, e.g., water, light, space, nutrients.
3. Because every population has genetic variation, some individuals are better able to get resources.
4. Those individuals that get resources reproduce and pass their genes on to the next generation; those that don't get as many resources, won't be as successful in reproducing, and will thus have fewer offspring making up the next generation. Thus, differential reproduction occurs.

Inquiring About Plants

How do new species evolve? One way is the following: a natural population, with genetic variation, is divided into two subpopulations because some seeds dispersed into a new habitat away from the rest of the plants. These two places have different environments. Because of the different environments, certain plants in one environment will be favored to reproduce and pass their genes on to the next generation. In the other environment, you might find that plants with different characteristics are favored. Genetic variation appears in both populations through mutations and the processes involved in sexual reproduction, such as random assortment, crossing over, and the random fertilization of gametes. Thus, over time, the plants in the two different populations gain collections of characteristics that are more suited to the specific environments in their specific habitat. If enough time and enough differences accumulate, plants from the two populations may not be able to reproduce when brought together—this is a sign that they are no longer part of the same species.

USING EVOLUTION THROUGHOUT YOUR COURSE

Now, consider specific major concepts in plant biology and how you might incorporate evolution into the study of those concepts—throughout your entire course.

When you teach about cells:

It is easy to incorporate information about endosymbiosis and how some prokaryotes evolved into eukaryotes. There are many different sources of information about this major evolutionary step, but here are some of the highlights. It is important to include the type of evidence that is used to establish and support the argument that endosymbiosis occurred. When you do so, you are "building a case." What is the evidence to support the claim that eukaryotes evolved in such a manner? Once students see the overwhelming set of evidence, their resistance to learning about the evolutionary history of life on Earth may be reduced.

Some evidence that endosymbiosis occurred:
- mitochondria and chloroplasts of eukaryotic cells are similar in size and morphology to prokaryotic cells
- mitochondria and chloroplasts reproduce, divide into two, in a manner similar to that of bacteria
- mitochondria and chloroplasts possess double membranes, and the inner one resembles bacterial membranes
- mitochondria and chloroplasts possess their own DNA (circular) and ribosomes similar to bacteria
- antibiotics that inhibit bacteria also inhibit mitochondrial and chloroplast protein synthesis

When you teach about structure (plant morphology):

There are many different examples of "form and function" in biology as a whole, as well as "plants are indicators of their environment." Both of these are related to the evolution of life on Earth.

"Form and function" is the generalization that is often observed in biology where the structure (form) of an object reveals or is related to its function. For instance, a spine is long, tapered, and pointed at one end (its form), and this form is directly related to its function of protection of the plant from an herbivore (as the animal approaches the plant to eat it, its nose or other facial part gets punctured with the pointed end of the spine). The herbivore quickly moves away from the plant, and the plant is protected from harm. Form and function applies to structures both small and large—from the shape of an enzyme (important for its function of interacting with specific substrates to catalyze a chemical reaction), to the form of a winged pine seed (important for dispersal away from the mother plant), to the form and function of an entire leaf (broad, which allows maximal surface area of the blade that can absorb sunlight, which is necessary for the leaf's photosynthetic function). Start getting your students to look at the form of objects and see how that form is related to function. Natural selection has selected for those forms by giving plants that possess the characteristic some advantage in a particular environment.

Another generalization in biology is that "plants are indicators of their environment." Plants are not motile (they cannot move from one place to another once they start growing somewhere). Whereas animals can move from one location to another if the temperature is too warm or the local food supply has greatly diminished, once a plant begins its life, it is stuck in the same place—for better or worse. A plant must be able to tolerate whatever the environment throws at it. It the plant does not possess the genetic information that would allow it to live, grow, and reproduce in a particular location, then the plant will die (or at least not reproduce), which means that its genes will not be passed to the next generation of plants. If, however, you find a plant growing and reproducing in a particular place, then it must be adapted to the surrounding environment—the temperature, rainfall, and soil in which it is growing. Thus, the plant is adapted to its environment. And....the plant can then be used to indicate what kind of environment in which it is living. For instance, if you find a succulent plant, one that has the ability to store water in its tissues, then you know that the environment in which this plant is growing is dry. Plants that are succulent are adapted to a dry environment and can beat the drought by storing water. Thus, if you find succulents growing in a particular place, those plants are indicating that there is limited amount of precipitation in their environment. Plants without the genetic information to produce succulent tissues have a limited ability to grow in deserts.

Inquiring About Plants

Both "form and function" and "plants are indictors of their environment" are evidence of evolution at work on plants and plant parts in different environments. For every concept you teach, is there something about the form of a particular object or is there something about the environment that can be determined by a plant growing there? (For instance, have students look at the plants seen in the scenery of movies to see if they can tell where the movie was shot.)

When you teach about plant tissues:

One can compare the water-conducting tissues of a diversity of plants—compare cells and tissues in species of plants that have been on Earth for a relatively long period of time compared to more recently-evolved plants, such as flowering plants. When one examines tracheids of pine trees, which are relatively ancient plants, these water-conducting cells are long and narrow, (which means they are like very thin straws) through which only small amounts of water may travel at any one time. What look like pores on the ends are only thin spots in the walls. On the other hand, the vessel elements of flowering plants are much shorter and broader and they have openings at both the top and the bottom and can be stacked on top of each other, which makes vessels like large, open-ended pipes through which much water may be transported (Sperry, 2003).

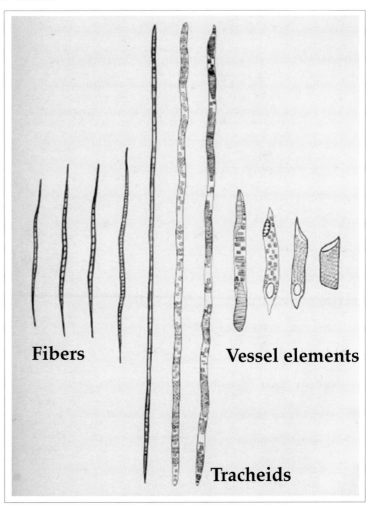

Figure 3.4 Examples of tracheids in relatively ancient plants and evolution of vessel elements in flowering plants. (From Bailey and Tupper, 1918)

The above example provides a good opportunity to talk about the misconception that evolution always produces the "best" way for something to happen and that evolution always helps organisms move toward some "goal" of "better." Natural selection may have favored those plants with larger water-carrying vessels. However, the thin pine tracheids still conduct water and "work" for the pines in the environment in which they are found—and pines still grow in many places around the world. In fact, the tallest living trees, the redwoods, have only tracheids to conduct water to their tops.

When you teach about plant physiology:
Ask, "Why aren't plants black?" The graph below shows both the absorption spectrum of chlorophyll (what colors of light does chlorophyll absorb?) and the action spectrum of photosynthesis of a typical plant (indicating that red and blue lights are absorbed and used in photosynthesis).

Figure 3.5 The absorption spectrum of chlorophyll a shows that chlorophyll a absorbs maximally at 430 and 662 nanometers. The rate of photosynthesis in different wavelengths of light is similar to the ability of chlorophyll a to absorb those wavelengths. The similarity of the action spectrum of photosynthesis and absorption spectrum of chlorophyll a suggests that chlorophyll a is the primary photosynthetic pigment and that light absorbed by chlorophyll a drives photosynthesis. (From Uno, Storey, and Moore, 2001)

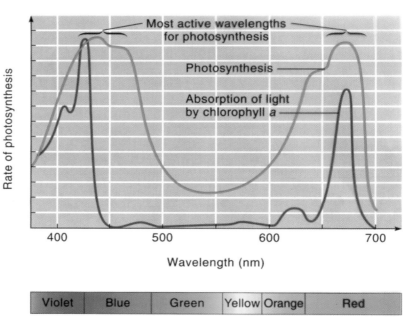

Plants are green, but, if light is necessary for photosynthesis, why aren't plants black? If they were black, plants could absorb and use all of the colors of the visible spectrum in photosynthesis. Green light is right in the middle of the visible light spectrum, and one would think that green light would be useful to

Inquiring About Plants

plants in photosynthesis, but it's really not. Chlorophyll reflects, but does not absorb, green light, which means it is not very useful in photosynthesis.

An answer to this conundrum might come from the evolutionary history of plants. Ancestors of plants evolved after the first photosynthetic organisms evolved, which were photosynthetic Eubacteria and Archaea. Early photosynthetic organisms used green light—so green light was not available to first ancestors of plants. Chlorophyll, the main photosynthetic pigment of plants, absorbs other colors. Some evidence for this evolutionary history comes from stromatolites (ancient fossils), in which we find evidence of photosynthetic bacteria, including those similar to the purple bacterium, *Halobacterium halobium*. Bacteriorhodopsin is a purple, photosynthetic pigment in *Halobacterium*. Bacteriorhodopsin absorbs broadly in the middle of the visible spectrum (absorbs green light). So, if purple bacteria and other photosynthetic organisms were prevalent in the waters before the ancestors of plants and they were already using green light as an energy source, green light was not readily available to the ancestors of plants because of the competition for this wavelength light. Another similar hypothesis has to do with species of Archaea, some of which use another light-absorbing molecule, retinal, to extract power from green wavelengths of light. If such green-light-absorbing organisms dominated the waters, this would have left only other "niches" for the ancestors of plants that could absorb wavelengths of sunlight other than green.

When you are teaching about plant reproductive biology:
How do we explain the great diversity of flowering plants (~250,000 species) compared to other organisms? There are a number of factors, however, the co-evolution of flowering plants with a wide variety of pollinators is part of the answer, and a wonderful example of evolution at work.

How do we explain the many different species of flowering plants? First, flowering plants can live on land—life evolved in the water, but those organisms that could move onto the land had many different habitats that could be colonized—but the issue is that water can be a limiting factor on land. Those organisms that possessed characteristics that allow them to collect, distribute and retain water have an advantage over those organisms that could not. Flowering plants have vascular tissue (xylem) that allows them to distribute water throughout their bodies, which means that they can grow tall and still get water to all of their parts. In addition, flowering plants have pollen grains that carry the sperm from one plant to another, so they can reproduce sexually more often than plants (such as ferns and mosses) that have swimming sperm and that can only reproduce sexually when there is water available in which their sperm can swim to the egg.

Also, it is important to note how quickly flowering plants can reproduce—certainly faster than gymnosperms, of which there are only a few hundred known species (compared to 250,000 species of flowering plants). Being able to reproduce quickly means that most flowering plants can have many more generations in the same amount of time than gymnosperms, which means that evolution can happen more quickly in flowering plants.

Finally, flowering plants have many different pollinators; they have co-evolved with a wide variety of animal pollinators. The following is a generalized description of the most common pollinators of flowering plants (the pollination agents), and the characteristics of flowers (syndromes) that are pollinated by those pollination agents (Table below and Figure 3.6). Note that the floral characteristics are those that are often associated with a particular group of pollination agents. However, there are, for example, thousands of different species of bees, and these insects have different preferences for floral attractants and rewards, which led to the evolution of thousands of species of bee-pollinated plants. This is what helped lead to a great diversity of flowering plants.

Table 3.1 Patterns of floral features associated with different pollination agents.

Pollination Agent	Flower Color	Odor	Flower Shape	Nectar	Other
Wind	Dull (green, browh)	None	Small sepals and petals Exposed stigma and stamen	None	Lots of pollen
Beetles	Dull or White	Strong, Fruity	Flat to bowl-shaped	None	Lots of pollen, Fleshy parts
Flies	Purple-brown	Strong, Foul	Flat or Trap blossom	Little	Often no food provided
Bees	Yellow, Blue, rarely Red	Sweet	Often a broad tube with landing platform	Present	Nectar guides
Butterflies	Pink is common	Strong	Deep, narrow tube	Present, not concealed	Flat, wide landing platform
Hummingbirds	Red is common	None	Tubular without landing platform	Abundant	Inferior ovary
Bats	White	Strong, Sweet	Large	Abundant	Lots of pollen

Inquiring About Plants

Figure. 3.6.
A sample of flowers illustrating some floral features linked to pollinator agents.

You might take students on a field trip and observe animal pollinators in action, noting the behavior of the animals as they move from flower to flower. Or, you could bring in a collection of flowers and, using the pollination syndrome chart, ask students to try to determine what the most likely pollination agent is for each species.

So, no matter what concept you are teaching, try to find some evolutionary component to the story you are building. Try to teach something about evolution for EVERY concept! This will help your students gain a deeper understanding of evolution as well as help them to organize the biological content that you are teaching.

Resources:
National Academy of Sciences—Thinking Evolutionarily
    http://nas-sites.org/thinkingevolutionarily/
University of California Museum of Paleontology—Understanding Evolution
    http://evolution.berkeley.edu

USING CLIMATE CHANGE TO MAKE CONNECTIONS ACROSS DISCIPLINES

In a nutshell: This activity provides historical climate data along with historic tree distributions along an elevation gradient and data on known environmental tolerances. Students work with the data to predict the movement of tree populations.

In the following, statements and questions for students are roman font, *typical student answers are in italic* type, and background information in sans serif.

→ The color patterns on the map below indicate the major biomes, recognizable regions of plant distribution, on Earth. What environmental factor might be responsible for the general pattern of the light blue and dark green colors in Canada and Russia? What is your evidence?

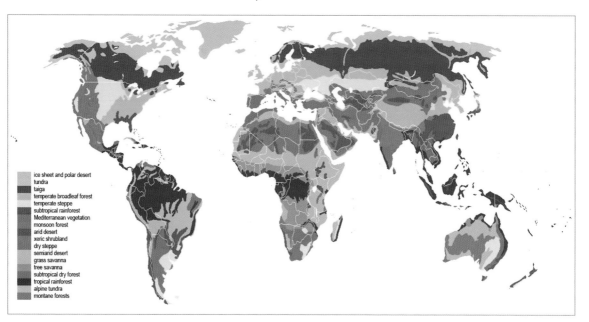

Legend:
- ice sheet and polar desert
- tundra
- taiga
- temperate broadleaf forest
- temperate steppe
- subtropical rainforest
- Mediterranean vegetation
- monsoon forest
- arid desert
- xeric shrubland
- dry steppe
- semiarid desert
- grass savanna
- tree savanna
- subtropical dry forest
- tropical rainforest
- alpine tundra
- montane forests

Figure 3.7 Major biomes of Earth.

Cold temperature is probably a major factor for these two regions, as well as shortness of the growing season and relatively little precipitation. They are the furthest north, thus farthest from the tropics.

Temperature is one of the environmental factors influencing plant distribution. In general, the further you move North or South of the equator (the higher the latitude), the cooler the temperature. In the US this trend runs from tropical Hawaii and Puerto Rico to arctic Alaska. The purple layer is called tundra and the green layer is taiga.

→ In North America, there is a patch of blue that extends from Canada north of Idaho and Montana almost to the Mexican border and then into Mexico. What might be responsible for this vertical North-South pattern of blue at right angles to the general horizontal pattern?

This is the Rocky Mountains.

Inquiring About Plants

⟐ Has anyone ever driven over a mountain pass like the Rocky Mountains on vacation? What happens to the temperature as you go up the mountain?

Temperature decreases as you go to higher elevations.

⟐ Alexander von Humboldt was the first scientist to recognize that the distribution of plants up a mountainside mirrors the distribution of plants at increasing latitude. Humboldt mapped the distribution of plants up and down a mountain in the Andes of Colombia, the tip of the finger of blue surrounded by dark green (rain forest) in South America on the map above.

What is the second main environmental factor associated with plant distribution? One hint is the pattern of the tropical rainforest (dark green) in the map above. Warm temperature is certainly involved because the horizontal pattern of dark green falls in the tropics on both sides of the equator. But light olive green, tropical grassland, is also in this band. What might the second factor be?

Rainfall is the second factor.

⟐ Precipitation is the second major factor determining plant geography and these two factors are related. What happens to air as it is heated (Hint: think hot air balloon).

Warm air rises.

⟐ In the tropics, the sun comes up at about the same time every day. Therefore it warms the air at about the same rate every day. Does warm air hold more or less water than cold air? (Hint: is it more or less humid in summer than in winter?)

Air is more humid in summer—warm air holds more water than cold air.

Figure 3.8 Dust cover of book describing von Humboldt's explorations on plant geography. (From Jackson, 2010)

⟐ In the humid tropics, as the air warms in the morning it picks up more moisture and rises. What will happen to the warm moist air when it gets to higher elevations?

The rising air cools, and as it does so water condenses.

The warm moist air cools as it gets higher and eventually water will start to condense, form clouds, and fall as rain—at about the same time every day in the tropical rainforest. The rising air also spreads, forming large circulation patterns as shown below.

The cool air sinks at about 30 degrees North and South of the equator. As it gets lower, the air begins to warm. Will it lose more moisture as it warms or will it pick up more moisture from the environment? In general deserts occur at about 30 degrees North and South of the equator because of down-draft of convection currents in the atmosphere.

→ Mountains also have a role to play in the distribution of precipitation. Look at the patch of brown (desert) west of the Rockies in the US, or across most of Northern Africa, on the map above. What geological feature covers Western Washington and Oregon and lies between California and Nevada?

Mountains are also on the west side of the desert.

There are also mountains on the West side of the desert, the Cascade Mountains and Coast Range in Washington and Oregon and the Coast Range and Sierra Nevada Mountains in California.

→ The prevailing winds in the northern hemisphere are primarily from the west. Will the incoming air on the West Coast be moist or dry after blowing over the Pacific Ocean? What will happen when they hit the west side of the mountains?

The incoming air will be humid from picking up water over the ocean. They will have to go up and over the mountains.

Moist incoming air will have to flow up and over the mountains. Just like in the tropics, as the air rises it will cool and moisture will condense producing rainfall. Coming down the other side, the now dry air will warm up and begin absorbing moisture from the surroundings producing a rain shadow on the down-wind side.

→ Knowing just temperature and precipitation, you can predict the dominant vegetation type of any place on Earth (Figure 3.9).

Below are average temperature and precipitation for several cities. What vegetation type would you expect?

Los Angeles, California: 21 °C; 38 cm

Desert

Juneau, Alaska: 3 °C; 205 cm

Temperate Forest

Fairbanks, Alaska: -9 °C; 30 cm

Tundra

↦ Scientists are concerned that global climate change is occurring and that the climate is generally warming. Below are Plant Hardiness Zone maps of the US from 1990 and 2012. What evidence do they provide that the climate is warming?

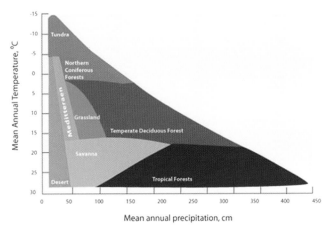

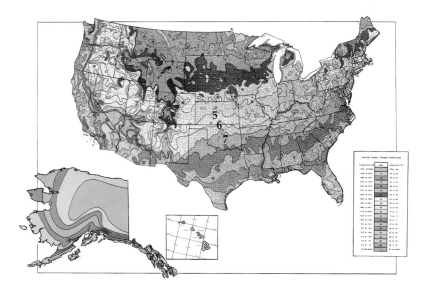

Figure 3.9 Relationship between precipitation, temperature, and vegetation type. (From Forseth, 2012)

Figure 3.10 USDA Plant Hardiness Zones in 1990 (bottom map) and 2012 (top map).

Hardiness zones are moving north.

Hardiness zones are migrating north. For instance, in 1990 zone 6 moved diagonally across Kansas covering nearly ½ the state. In 2012 zone 6 (the two darker shades of green) cover nearly all of the state and zone 7 touches the southern border.

•▸ Given what you know about the distribution of plants on mountains, what do you predict will happen to plant distribution on mountains as climate changes?

Plant populations should be moving to higher elevations on mountains.

•▸ An international group of researchers, the Global Observation Research Initiative in Alpine Environments (GLORIA) has established a network of research on mountain tops around the world as shown below.

Figure 3.11 International network of research on mountain tops. (From Pauli et al., 2012)

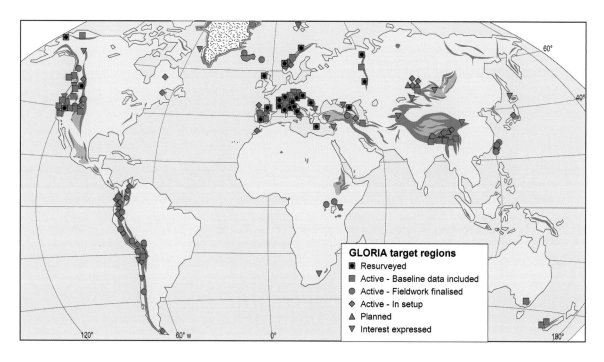

At each site, the same kinds of measurements are made at different elevations. At each site samples begin in the zone above tree line and at three sequentially lower zones through the alpine forests.

Inquiring About Plants

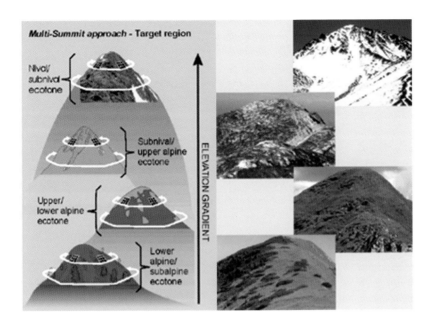

Figure 3.12 Standardized sampling locations for GLORIA mountain top studies. Four mountains of different heights are sampled at each study location. (From Global Observation Research Initiative in Alpine Environments)

Figure 3.13 Photographs taken at a similar location near Lewiston, Montana, in 1917 and 1959. (From Philips, 1963)

Explain how the photographs do or do not support climate change.

The top image in Figure 3.13 was from 1917, the bottom is the same view in Montana from 1959.

There are now many more trees in the distant river valley in front of the mountains, and also moving up the slopes.

→ The map below, from a 2012 *Science* paper by Pauli and colleagues, summarizes data from all of the GLORIA sites in Europe. At each site are two yellow bars indicating the total number of vascular plants observed at that site in 2001 (left) and 2008 (right). The number of endemic species (native to a particular area) found only at that site are indicated by the red portions of the bars.

What has happened to the number of species at almost every site during the past seven years (2001–2008)?

The total number of species has mostly gone up but the number of endemic species varied.

→ Explain how the data do or do not support climate change.

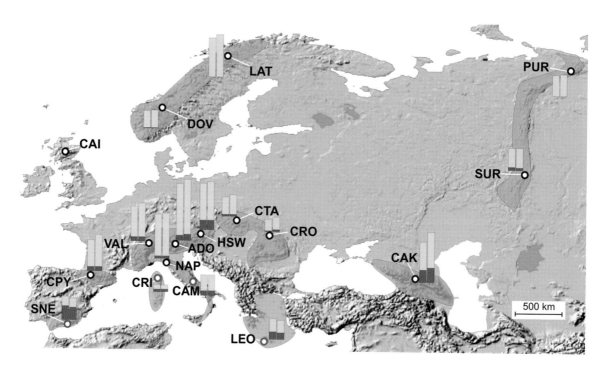

Figure 3.14 GLORIA study sites at various mountain ranges (blue) in Europe. Bars indicate total number of species sampled in 2001 (left) and 2008 (right). Red indicates endemic species. (From Pauli et al., 2012)

The data support climate change. As global temperatures increase plant species are migrating up the mountains.

→ The graph and table in Figure 3.15 are more specific data for the NAP site in Northern Italy. Which of the mountains, A, B, C, or D has the highest elevation?

D

D is 2893m tall.

→ Can you explain why D has the fewest species?

Colder temperatures at higher elevations limit plant growth. For instance, there are no tree species at site D.

→ Explain how the data in the graph and table does or does not support climate change. Of course, the foreseeable problem is that if plants have to migrate up the mountains, they will eventually reach the top and then go extinct from that

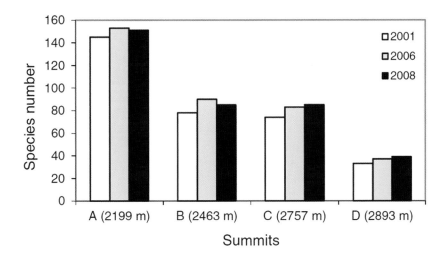

Figure 3.15 Change in numbers of mountain-top plant species from 2001-2008, with change in number of conifer species at the four summits sampled shown below. (From Erschbamer et al., 2011)

Numbers of young trees on summits (A–D) in 2001, 2006 and 2008.

Summit	Species	2001	2006	2008
A	*Larix decidua*	0	3	5
	Picea abies	1	2	3
	Pinus cembra	3	5	9
	Sorbus aucuparia	0	0	1
B	*Larix decidua*	0	1	3
C	*Larix decidua*	0	1	2
D	no tree species			

The data support climate change: higher mountains have fewer species, but the number of species is increasing over the 7-year period on all mountains.

Similarly, and like the photographs from Montana shown above, the number of trees is also increasing over time.

The final set of data, from a 2011 article by Crimmins and colleagues in *Science*, are from the Coast Range and Sierra Nevada range of California as shown on the map (Figure 3.16).

In the data below (Table 3.2), change in elevation is recorded for 73 plant species found in the mountains. If the change was a statistically significant difference, the number is highlighted.

Do these data support that global warming is responsible for plants moving to higher elevations on mountains?

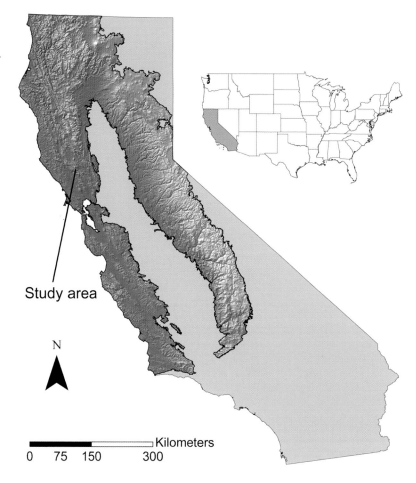

Figure 3.16 Map of California showing mountain ranges studied. (From Crimmins et al., 2011)

Study area

N

|————————————————| Kilometers
0   75   150   300

NO! While the distribution of some species has gone up, the distribution of other species has gone down.

The data do NOT support that global warming is responsible for plants migrating up mountains. For instance, both Juniper species at the bottom of the first column have migrated DOWN the mountainside—in one case by 745 m—the most of any plant. *Pteridium aquifolium*, a fern, migrated up the mountain by about the same distance.

▸▸ If some species migrated down the mountain, instead of up, what other factor associated with climate change might have been involved?

Precipitation might be a factor in migration.

Table 3.2 Elevation change, positive or negative, in meters. (From Crimmins et al., 2011)

Species	Elevation	Species	Elevation
Abies concolor	-25.1	Lithocarpus densiflorus	-492.7
Abies magnifica	147.0	Lotus scoparius	-
Adenostoma fasciculatum	-12.9	Monardella odoratissima	-
Aesculus californica	-433.6	Pinus albicaulis	-
Amelanchier alnifolia	-276.8	Pinus attenuata	284.2
Artemisia californica	-	Pinus jeffreyi	86.1
Arctostaphylos glauca	41.4	Pinus lambertiana	-176.8
Arbutus menziesii	234.3	Pinus monticola	-416.3
Arctostaphylos nevadensis	-197.0	Pinus ponderosa	70.8
Artemisia tridenta	-253.7	Pinus sabiniana	-119.7
Arctostaphylos viscida	-100.8	Prunus emarginata	-36.7
Baccharis pilularis	-	Pseudotsuga menziesii	239.0
Calocedrus decurrens	-80.0	Pteridium aquilinum	744.2
Ceanothus cordulatus	214.4	Purshia tridentata	-316.3
Ceanothus cuneatus	-129.9	Quercus agrifolia	-
Ceanothus integerrimus	-160.6	Quercus berberidifolia	667.7
Cercocarpus ledifolius	-711.7	Quercus chrysolepis	-94.9
Cercocarpus montanus	-76.0	Quercus douglasii	-30.4
Ceanothus prostratus	-20.6	Quercus durata	-156.8
Ceanothus velutinus	-270.7	Quercus garryana	-31.0
Chrysolepis chrysophylla	-439.7	Quercus kelloggii	-37.0
Chamaebatia foliolosa	23.7	Quercus lobata	-
Chrysolepis sempervirens	211.2	Quercus vacciniifolia	-93.5
Corylus cornuta	-276.4	Quercus wislizeni	-
Cornus nuttallii	-285.3	Rhamnus crocea	483.6
Eriodictyon californicum	-61.5	Rhamnus ilicifolia	-232.3
Eriogonum fasciculatum	61.4	Rhus trilobata	160.2
Ericameria linearifolia	-59.9	Ribes roezlii	247.0
Ericameria nauseosa	-313.9	Salvia leucophylla	293.6
Frangula californica	-	Symphoricarpos mollis	-690.7
Fraxinus dipetala	-100.4	Toxicodendron diversilobum	162.7
Garrya fremontii	-154.4	Tsuga mertensiana	-255.1
Heteromeles arbutifolia	-53.1	Umbellularia californica	-558.2
Hesperoyucca whipplei	-130.6	Vaccinium ovatum	-186.8
Juniperus californica	-151.5	Vaccinium parvifolium	-206.9
Juniperus occidentalis	-745.1	Vulpia myuros	-13.7
		Wyethia mollis	-352.6

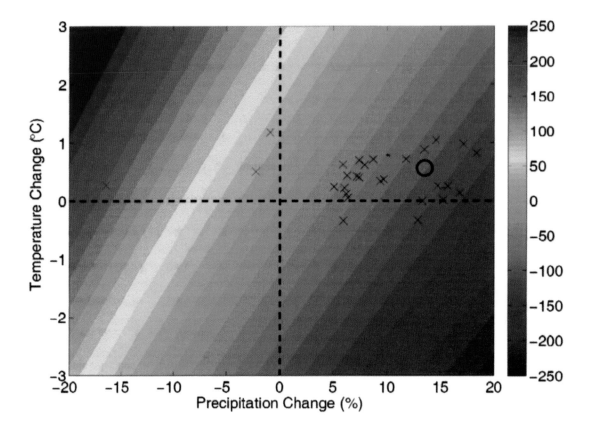

Figure 3.17 Relationship of temperature change and precipitation change to elevation change of mountain plant species. (From Crimmins et al., 2011)

The response of a plant to climate change will depend both on its response to changing temperature conditions (Left Y-axis) AND response to changing precipitation (X-axis). Whichever variable is more critical for a particular species will determine if it migrates up or down the mountain (Right Y-axis).

Do you think that all plant species are able to migrate up or down a mountain (or move North or South) from their present range? What characteristics of a plant species might prevent it from being able to respond to changes in the environment? (Some plant species may have very long life histories, and therefore not be able to migrate before changes in the climate are fatal to the species.)

Resources:
National Arbor Day Hardiness Zone Maps
http://www.arborday.org/media/highresolution.cfm

Global Observation Research Initiative in Alpine Environments (GLORIA)
http://www.gloria.ac.at

Inquiring About Plants

3.2. Putting Practices into Action in Student-centered Open Investigations

The opportunity to ask their own questions to investigate is highly motivating to many students. This ownership and motivation feeds into students taking responsibility for their own learning. On the other hand, open-ended investigations can also be intimidating to students as they plan their own investigation for the first time. The student handout below provides some supporting structure to guide their actions. It provides examples to help students generate research ideas and questions about germination and plant growth.

BACKROUND

How can you tell if your houseplants need fertilizer? How well would your houseplants grow if you added twice the recommended amount of fertilizer to their pot or to your plants in your garden? What plants are the birds eating in your backyard, and are they eating seeds, fruits, or insects? How hot does the soil get in the sun next to a tree compared to the soil in the shade, and how does this affect root growth? These are questions that you might ask about everyday phenomena you see in the world around you. But how might you answer these questions? The independent investigation is your opportunity to study some small part of your natural world and to answer one of your own questions.

There are four main steps to your project:

Select and outline a research problem.
Design an investigation.
Conduct the investigation.
Write the project report.

Your investigation can be either an experiment or an observational study. In an experiment you will manipulate part of the environment you are studying to answer a question. This manipulation may include a setup with control and variable conditions. In an observational study you will not change the conditions of the environment nor create controlled and experimental situations. For instance, if you observe the kind and number of individual insects that visit the flowers of a particular species of plant in a field near your house, you are doing an observational study. However, you are doing an experiment if you remove petals from some flowers and not from others of the same plant species to determine if insects are attracted to the flowers because of their petals.

Either kind of study—observational or experimental—is satisfactory for this project. Whichever you choose, there are several skills you will be asked to develop and use. The first skill is to formulate a research problem that you can answer based on your observations of the world. You are not expected to win a Nobel prize with your work. It is important, however, that the project be well designed and conducted properly. Once you have an idea for the research problem you would like to study, you need to decide on the method of investigation. You must be able to complete your project with limited materials and time. Set up your study and then begin collecting data systematically and thoroughly. The data you collect will be processed, interpreted, and then incorporated into a write-up of the results of your investigation. The final skill involved is the writing of the report.

SELECTING AND OUTLINING A RESEARCH PROBLEM

First, choose an area of interest. Do you like to work indoors or outdoors? Do you like anatomy, or physiology, or plant-animal interactions? Do you like to work with flowers or with leaves? Next, observe and ask questions about a situation related to your area of interest. Select one question and form your research problem based on this question. A good research problem is simple, specific, and feasible in regard to the time and materials available for the study. The entire project will be based on your observations.

There is an infinite number of projects from which you could choose. Consider just the soils in your community. Do all kinds of soil absorb rainfall at the same rate? Do all the soils hold the same amount of water? Do all soils have the same kind of acidity? Is the amount of water that can be held related to any particular characteristic of the soil? Does soil type make any difference in plant growth? Does the amount of light or depth you plant your seeds affect plant growth? What happens to the size of individual plants if there are crowded in a pot? All these questions can lead to different independent investigations.

If no project immediately comes to mind, think about what affects your health—either positive or negative factors. For instance, do you take vitamins because you think they keep you healthy? What would be the effect on plant growth if you added vitamins to the soil? Would the plants be healthier, and how could you tell if they were? (Remember, if you see no differences between your control and variable, you still have a valid project!)

Consider the following research questions. Some of them are workable and some of them are not.

Question 1: What kinds of seeds are found in the soil underneath the canopy of maple trees—are only maple seeds found here?

This question is acceptable—it is simple, specific, and can be completed with your background, materials, and time. This is an observational study.

Question 2: What will happen if I play soothing music versus rock and roll to my plants—will plants grown in rock and roll music be as healthy as plants grown in soothing music?

This question is unacceptable. All aspects of an experimental study must be the same except for the one variable being tested. This means that the plants exposed to the different kinds of music must be kept in identical, but separate, rooms with identical amounts of sunlight, identical temperatures, identical humidity, and cared for in an identical manner. This is an impossible task. If you kept plants in the same room and then moved them into other rooms just to play music to them, you create other variables that destroy the precision of the experiments (disturbing the soil as you move the plants, leaving plants in the sun for different lengths of time, etc.).

Question 3: What are those birds doing in my oak tree?

This is a good starting question for an observational study, but it needs to be focused. Which birds are you talking about? Are the birds there for food, or shelter, or for rest, and how could you determine this?

DESIGNING YOUR INVESTIGATION

You will turn in a proposal for your investigation. Your proposal should focus on what you intend to study and how you will go about it. It should begin with an introduction that includes the general problem, background information, and question you are attempting to answer. Following this, you should have a methods statement with a detailed outline of the materials and equipment needed to complete your project. In this section you should also include a description of each of the methods that you intend to use. Can other students read this section and set up an appropriate investigation just based on your description of the materials and methods you will plan to use?

Use This Checklist to Evaluate Your Proposal

Introduction

_____Did you state the specific question you will study?

_____What is your rationale for choosing the problem? Why did you choose this particular project; what first made you think of this project?

Materials and Methods

_____Where are you going to work?

_____How are you going to investigate the problem? Give detailed descriptions of your methods.

_____What are you going to measure (height, number, size?) and how often are you going to measure or sample?

_____Do you have enough samples to allow you to average your results?

_____What materials and equipment do you need to complete your research?

_____Will the data you are going to collect actually let you answer the question you have asked?

_____Are your control and variable(s) setups clearly defined?

_____Does your schedule allow you enough time to complete the project and to write up the report?

CONDUCTING YOUR INVESTIGATION

Keep a notebook with all of your observations and measurements. Make certain that you write everything down—do not think you will remember the information. Date each entry and include time of observations and other references (which pot or plant you are observing). If appropriate, you should average your data and graph the data for easy interpretation. Remember there is a difference between raw data (numbers that you have collected throughout the investigation) and processed data (the same numbers totaled and averaged).

WRITING THE FINAL REPORT

There should be four sections to your report: (1) Introduction; (2) Materials and Methods; (3) Results; (4) Discussion.

The Introduction and Materials and Methods sections should be similar to what you included in the proposal. However, you should include any changes that resulted from conducting the project.

The Results section should begin with an overall picture of all the results that you have found and the general trends that have shown up (e.g., birds ate more red berries than blue berries during the morning hours). In this section, you should include your tables and graphs, but not your pages of notes you kept throughout the project (attach these to the end of the report). Avoid analyzing the significance or apparent meaning of the data in this section—leave that for the Discussion.

In the final section, the Discussion, you should refer back to the data that you present in the Results section, but do not repeat the data. Avoid making unjustified conclusions—keep your conclusions to the experiment you completed. Do not try to make broad, all-encompassing generalizations based on your results. For instance, if you find that lettuce seeds germinate at a faster rate in red light than any other color, do not generalize that all kinds of seeds will germinate at the greatest rate in red light. Also do not make unjustified speculations, e.g., because the lettuce seeds germinated at a faster rate in the red light, lettuce plants will grow better in red light than in sunlight.

Use This Checklist to Help Organize Your Report

_____Does your title indicate the content of the report?

_____Have you clearly stated the problem or question you have studied?

_____Have all your methods been described concisely and accurately?

_____Have measurements or observations been made adequately for all variables in your project?

_____Have you omitted all interpretations and conclusions from the results section?

_____Have all the results related to the research problem been reported in a concise and understandable manner?

_____Have the data been processed in a useful form (graphs or tables) and have you left your raw data out?

_____Are your graphs, pictures, and tables drawn and labeled clearly and properly?

_____In your discussion, have all of your reported results been interpreted and discussed in relation to your original question?

_____Have you stated your conclusions based on your results?

_____Have you pointed out possible errors or biases?

_____Have you written all of your sections in the most concise manner possible?

This is not a library report. In fact, you don't need to use any text during the project. This is an investigation on some aspect of plant life, and your questions and observations are the most important parts of it. You can choose any topic related to botany, but there are some restrictions:

1. Make certain that you can get the equipment or materials to complete your project. We will supply pots, soil, seeds (peas, beans, corn, radish, or sunflower), fertilizer, and a place to grow your plants. You may, however, want to keep your plants at home. Anything "out of the ordinary" must be provided by you. Check with us before you buy anything—we may have it.

2. Make certain that you will have time to complete the project. For instance, do not study the life cycle of an oak tree.

3. Creativity will be rewarded, however, it is not essential. What is important is that you make many careful observations while your plants are growing and that you think about your results and what they could mean. Measurements are part of your complete set of observations. When selecting a project, do not be too ambitious.

4. You must submit a one-page proposal about your project on the deadline given in class. It should include: a) what you are going to do, b) what materials are needed, c) where you are going to keep your plants, d) a brief schedule of completion, and how you are going to set up the project. The Final Report of the project should be 3–5 pages and turned in on the due date given in class.

- Important points to remember:
- You must grow a mininum of 15 plants.
- You must measure at least *twice* a week.
- You must have the *same number* of plants growing in each pot (3–5)—unless plants die due to the treatment or care they receive.
- Do not fertilize your soil (unless that is part of your experiment).
- If you are going to add anything to the soil other than water, use vermiculite.
- Keep your pots in a warm, brightly-lit place (south-facing window).
- Water as necessary—test the soil with your finger. If the soil is moist at 1 inch below the soil, then you don't need to add water.
- Plant seeds *no more* than ½ inches below the soil surface.
- In most cases, start with seeds, not whole plants—that's because you need to be sure that all of your plants are the same age.

- Start your project as soon as it has been okayed because plants may die, and then you'll need to start over.

EXAMPLES OF INDEPENDENT PROJECTS

You don't have to choose one of these—be creative. Go outside, take a walk in the woods, and make observations of how plants are growing there. Look in your text or look at plants to generate an idea. Ask yourself what affects the growth and development of plants. Do not do a grafting project. Do not do a project about growing plants in the light and dark. Do not do a project about talking to your plants or playing music to your plants.

Some ideas to think about:

- How will crowding affect plant growth?
- Compare the growth of plants in:
  - Different types of soils
  - Different light intensities
  - Different colors of light
  - Different amounts of moisture in the soil
  - Different kinds of water or water solution
- What will happen to plants grown with different kinds and/or concentrations of?
  - Herbicide
  - Fertilizer
  - Individual nutrients such as nitrogen, phosphorus, or potassium
  - Plant hormones such as auxin, gibberellin, or ethylene
- How will the following affect plant growth?
  - Sugar
  - Salt
  - Caffeine
  - Vitamins
  - Pollutants
  - Soil pH
  - Temperature
  - Acid rain
  - Humidity
  - Damage to the leaves

- What happens to the size of the root system if leaves are cut off a plant?
- What is the effect of grazing (or clipping) on the growth of a grass plant?
- Are there more seedlings of a tree nearby the mother plant, or at some distance away from it?
- Do larger fruits contain more seeds than smaller fruits of the same species?
- Do larger seeds germinate and grow faster than smaller ones?
- Are there more plants of a certain kind in the sun or the shade?
- Do all plants of a certain kind and age produce the same number of seeds and the same size of seeds?
- How far do seeds of different sizes travel from the mother plant?
- What kind or color of seeds or fruits do birds prefer to eat?
- Are there more algae or water plants growing in standing water or in flowing water?
- Do certain kinds of insects visit a particular flower? And for how long? What are they doing at the flower?

There are an infinite number of possible projects! As you complete other investigations in your course, keep a log of the questions you asked along the way. Perhaps one of these questions can be used as a start of your independent project. Look at some plants growing inside or outside, and ask some questions about things you see. Leaf through a plant biology text and see if you can gain any ideas from the material there. Use your imagination, but remember to choose a project that asks just one question.

THE CELERY CHALLENGE AND CONNECTING STUDENT TEAMS TO PLANTINGSCIENCE MENTORS

In a nutshell: This investigation is framed as a challenge for student teams to cause and explain the most extreme bending in celery stalks. Advantages are that the investigation taps into student motivation and encourages explanations using evidence from multiple sources. The aim here is for students to generate working models of how structure-function relationships and environmental conditions influence water movement into and through a plant. Students build on observations from two guided investigations to design their own team experiments and form explanations about celery bending integrating concepts of osmosis, transpiration, and cell types.

As you plan the sequence of investigations for your course, you may choose to include some investigations carried out as independent student projects, some done as a class, and others conducted by teams. There are benefits to involving students in all these types of investigations, and a variety of ways to include expectations that students make evidence-based claims in communicating their research to others. If you want numerous informal opportunities for your students to describe ideas and reasoning in their own words, then perhaps build on the collaborative learning that occurs in team projects. A way to further enhance the collaboration and communication in investigations is to connect to today's digital science learning environments. Your options are vast—limited primarily only by your schools' access to high speed internet and computers.

One such opportunity is the PlantingScience program (www.PlantingScience. org). It is an award-winning online learning community where students are engaged as research teams to design and carry out inquiries about plants while scientists mentor them through the experience. Student teams conduct the plant investigations in their classrooms, and the learning extends beyond the school walls in online discussions with scientists who help students think and work like scientists. Talking online with a scientist and taking ownership of their own research question is exciting and motivating to students. Asynchronous discussion boards for each student research team and their mentor make student thinking visible. Professionals from more than fourteen scientific organizations have volunteered as online mentors to middle and high school student teams, offering encouragement, guidance, and probing questions to help students reflect and dig deeper into ideas. Mentors also serve an important role welcoming and acculturating students to the scientific enterprise.

The Celery Challenge (and the Corn Competition) is part of a suite of modules available in PlantingScience covering big ideas in biology that can be approached in the classroom through first-hand plant investigations. Ecological interactions, inheritance and variation, growth and development, energy flow, structure and function are encompassed in this suite. As you might expect, the curricular modules present the concepts integrated with skills for participating productively in science. So, you will find they link content and practices of science in keeping with the Next Generation Science Standards (http://www.nextgenscience.org). The PlantingScience curricular resources are freely available for anyone to access online. We invite you to download the complete teacher and student guides and mentor tips for the Celery Challenge, which have background information, sample 12-day learning sequences, technical notes, real-world connections, and resources. The abbreviated teacher version below highlights how initial guided explorations of plant tissues and cells prepare students to develop an understanding of structure-function relationships that teams later use in open inquiry about celery bending.

This investigation uses inexpensive materials readily available from the grocery store for students to ask and answer relatively sophisticated questions about water movement in plants. Be sure to use the celery stalks with leaves attached, not the trimmed "celery hearts." The culminating grand challenge is for student teams to create the most bending in celery and to describe the explanatory mechanisms for what they observed and tested. Before teams tackle this seemingly simple task—which actually draws on integrating understandings of cell types, osmosis, and transpiration—set the stage with preparatory activities.

Begin with a class discussion on the topic of how plants move water. Ask, for example, "What is happening when a plant wilts?, What makes young stems and leaf petioles stiff enough to stand upright or hold a leaf towards light?, What is happening when a Venus flytrap opens or closes?" These are some of the juicy questions to engage students' interest in the topic and explore their prior knowledge in the Celery Challenge teacher's guide available at www.PlantingScience. org.

 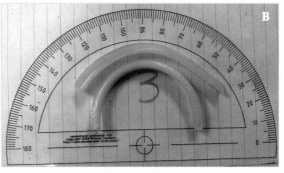

Figure 3.18 Examples of student work in the Celery Challenge.

A scenario about preparing celery stick appetizers in advance of a party sets up the first guided investigation with a real world context. Here students observe osmosis and how it influences the firmness of celery cells and tissues. Students measure, weigh, and make qualitative descriptions of celery stalks before and after soaking it overnight in salt solutions that are hypotonic or hypertonic relative to living cells. After students record their data and team members calculate averages for each treatment, have students summarize their ideas about the treatments and data collection techniques in their lab notebooks by answering: (1) Which treatment produced the most bending, and the least? (2) Was there any difficulty in determining which treatment caused the most bending? If so, describe why. (3) Did any other characteristics clearly differ among the treatments? Note these as well. (4) What questions to you have about the changes you observed? List at least five, include at least one related to the qualitative data and at least one to your quantitative data.

Depending on the salt concentrations students use, salt treatments may be hypertonic, isotonic, or hypotonic relative to celery cells. So it is possible that bending will not occur. Students will likely notice that the hypertonic salt treatment makes the celery flexible and rubbery, but they might not immediately notice the reduction in dimensions and overall mass. Tap water is hypotonic relative to living cells, so celery stalks soaked in it will swell, increasing their dimensions, mass, and firmness.

In the second guided activity, students observe transpiration and different tissue types as they examine how a colored solution moves through a celery stalk. As with the previous activity, have students record qualitative observations before and after standing the stalks in the colored liquid overnight. Students then focus on finding evidence of dye by examining the intact exterior and cross-sections made at various places along the length of the celery stalk.

Students will typically notice the submerged base of the celery stalk is stained, but the color is darkest in a series of spots just inside the ridges on the outside of the celery stalk (Note how the students should be able to describe or draw what they see even if the cell type vocabulary is not on the tip of their tongues). Recalling the previous activity may help students grasp that some amount of osmosis occurred in the immersed cells due to the hypotonicity of the liquid. Using a dissecting microscope, students will see the darker spots are associated with the open ends of the xylem tubes. And examining cross-sections of the celery stalk above the liquid level, students will see only xylem is stained. Osmosis is limited to the immersed cells, while transpiration and the vascular system move most of the water, and therefore the dye, up the stalk.

Be sure to also include some time and materials for students' open exploration of celery anatomy and physiology. This is a chance to "muck around" and follow up on observations and ideas, such as "What do cross sections of celery stalks soaked in different salt concentrations look like" or "What would happen if we used only an inner portion of celery?" Students often have an idea that the outer, ridged side of the celery is less flexible than its inner side but may not have ideas how that relates to bending, particularly if they are unaware that plant cell types differ in flexibility. Providing this time for teams to try out some brainstorming ideas and techniques before teams are asked to generate a testable question will go a long way to improving the quality of their research and giving students confidence in skills working as scientists.

While all teams will broadly be asking "How far can celery tissues bend" and "What conditions create the most bending," have each team select a narrower, testable research question. Within the framework of the culminating challenge, student teams can ask a broad range investigations, such as the role of xylem and other tissues in stiffening, the effect of osmosis, the effect of different tissue types, the effect of cross-sectional shape, age and position of the celery section, the rate of transport in the xylem and flow through stomata and role of environmental influences of temperature, humidity, light, or air movement.

Teams doing the Celery Challenge as part of PlantingScience receive feedback about their ideas for research questions and experimental designs as they carry out the culminating open-ended investigation. Teams also benefit from a mentor's guidance as they consider the mechanisms that explain their data. To allow teams to compare findings of their various studies and synthesize student ideas to answer the big question of what makes celery bend, close with a class discussion.

DEVELOPING YOUR OWN INQUIRY-BASED ACTIVITIES

Now that you have sampled a few ideas to increase the level of student engagement, student learning, and student-centered activities in your class, you might try to develop your own inquiry-based activities. There are many different ways to engage students and to help them learn about plant biology, but there are a few characteristics of good activities. Choose a topic/concept you want to teach, and then focus on how you can help your students learn about that subject matter. As you develop your activity around some biological content, think about what you hope the activity will also do. . . . such as uncover misconceptions, or get students to work with datasets, or to give them an opportunity to design and conduct their own experiment. One question to always ask is: am I telling students something that they could discover on their own (with a little guidance or help from me)?

Here are some components of a pedagogically sound activity. All activities should incorporate one or more of these components. But no single activity will be able to accomplish everything on this list at the same time.

A CHECKLIST OF COMPONENTS FOR STUDENT ACTIVITIES— ACTIVITIES SHOULD:

1. Uncover misconceptions and naïve explanations that students possess about a concept (once you know what the misconceptions are, how will you help your students gain more appropriate and accurate understandings?)
2. Generate and hold student interest (activities that are relevant to the lives of students or bizarre are more appealing—for instance, using examples of common foods or poisonous plants might be

more interesting than that of a tropical plant that students will never see for themselves).

3. **Allow students to be active learners** (instead of lecturing to them, can students discover some information on their own through the activity itself?)

4. **Help students gain a deeper understanding of concepts** (the more time students spend actively engaged in learning about a topic, the greater the chance they will understand and be able to apply their knowledge. For more difficult concepts, complementary activities that stress different thinking skills and get at the same material from different points of view would help.)

5. **Let students learn selected content in context** (for example, can they learn about transpiration by conducting an investigation on the relationship between the surface area of leaves and the rate of water loss—as opposed to you telling them about transpiration and the relationship?)

6. **Allow students to experience science as a process** (consider how many laboratory/field investigations you incorporate into your course. Are they integrated with the "lecture" part of your course? Can you include an investigative component for every major concept you teach?)

7. **Provide opportunities for students to make careful observations of natural phenomena** (good investigations begin with careful observations—get students to spend time and to write down all of their observations and questions about some biological phenomenon or process.)

8. **Encourage students to ask questions and to describe how they might answer their own questions** (the more time they spend observing, the more questions that will arise—but you have to constantly encourage your students to ask questions.)

9. **Provide opportunities for students to design experiments and predict results** (allow students to design and conduct an experiment to answer one of their own questions)

10. **Provide opportunities for students to process and interpret data and information** (data can be collected by students and then analyzed for trends, patterns, or significance, or you can use datasets that others have collected and reported and use those as a basis for student analysis)

11. **Encourage students to collaborate and give them a forum to communicate** (while student group work can be beneficial, what did the individual student contribute to the group's work, and

what does that individual student understand about the topic/
investigation? Use a variety of different ways that students may
communicate—in discussions, oral presentations, poster ses-
sions, lab reports. A sample miniposter is available from National
Association of Biology Teachers website at http://www.nabt.org/
blog/wp-content/uploads/2010/04/Sample-miniposter.pdf)

12. **Help students make connections between information, concepts, and themes** (the use of themes can help students connect
all of the information they learn, but how does one concept connect to another? Always ask students to connect what they are
learning about to what they have already learned.)

13. **Promote student reflection on their understanding** (formative
feedback can help students understand what they know and don't
know—and how they might gain a better understanding of the
material)

14. **Assess deep student understanding** (ask more questions that
require students to explain concepts in their own words, or to
apply their knowledge to a new situation that they have never
heard about, or to teach a subject to the class—not just report
about it, but teach their peers)

Resources

Online materials are included with particular activities, but here is a brief list of resources that might be generally helpful in teaching biology, and plant biology.

PlantingScience
http://www.plantingscience.org

Plants-in-Motion
http://plantsinmotion.bio.indiana.edu/plantmotion/starthere.html

Core Concepts in Plant Biology
http://c.ymcdn.com/sites/my.aspb.org/resource/resmgr/Education/
    Undergradplantbio_conceptsan.pdf

Science and Plants for Schools
http://www.saps.org.uk

College Board Advanced Placement (AP) Biology Course
http://apcentral.collegeboard.com/apc/public/courses/teachers_cor-
    ner/2117.html

Next Generation Science Standards
http://www.nextgenscience.org

Understanding Science
http://undsci.berkeley.edu

Vision and Change in Undergraduate Biology Education
http://visionandchange.org

Introductory Biology Project
http://ibp.ou.edu

NatureScitable
http://www.nature.com/scitable

References

CHAPTER 1

Becklin, K. 2008. A Co-evolutionary Arms Race: Understanding Plant-Herbivore Interactions. *American Biology Teacher* 70: 288–292.

Chamovitz, D. 2012. *What a Plant Knows: A Field Guide to the Senses*. Scientific American/Farrar, Straus and Giroux, New York.

Chase, M.F., M.J.M. Christenhusz, D. Sanders, and M.F. Fay. 2009. Murderous plants: Victorian Gothic, Darwin and modern insights into vegetable carnivory. *Botanical Journal of the Linnean Society* 161: 329–356.

Darwin, Charles. 1881. *The Power of Movement in Plants*. New York: D. Appleton and Co.

Dittmer, H. 1937. A Qualitative Study of the Roots and Root Hairs of a Winter Rye Plant (*Secale cereale*). *American Journal of Botany* 24(7): 417-420.

Dayton, L. 2001. Philodendrons Like it Hot and Heavy. *Science* Now 11 April 2001.

Klein, R.M. and P.C. Edsall. 1965. On the Reported Effects of Sound on the Growth of Plants. *BioScience* 15: 125–126.

Raloff, J. 2000. Fighting Cancer from the Cabbage Patch. *Science News* 158(13): 198.

Tribe, Michael A. and Derek Peacock. 1976. *Basic Biology Course, Unit 2, Organisms and their Environment, Book 4: Ecology Game*. Cambridge: Cambridge University Press.

Wagner, G.J., E. Wang, and R.W. Shepherd. 2004. New Approaches for Studying and Exploiting an Old Protuberance, the Plant Trichome. *Annals of Botany* 93: 3–11.

Woodward, F.I. and C.K. Kelly. 1995. The Influence of CO_2 Concentration on Stomatal Density. *New Phytologist* 131: 311-327.

CHAPTER 2

Advanced Placement Biology Curriculum Framework, 2012–2013. 2011. New York: The College Board.

Ainsworth, S., V. Prain, and R. Tytler. 2011. Drawing to Learn In Science. *Science* 333: 1096–1097.

Baum, D.A. and S.D. Smith. 2012. *Tree-thinking: An Introduction to Phylogenetic Biology*. Roberts and Company. Greenwood Village, Colorado.

Beal, William J. 1880. The New Botany. *Transactions of the Michigan State Teachers' Association at the Thirtieth Annual Meeting, held at Lansing, December 28, 29, and 30, 1880.*

Costenson, Kenneth and Anton Lawson. 1986. Why Isn't Inquiry Used in More the Classroom? *American Biology Teacher* 48(3): 150–158.

D'Avanzo, C. 2008. Biology Concept Inventories: Overview, Status, and Next Steps. *BioScience* 58: 1–7.

Ebert-May, D., J. Batzli, and H. Lim. 2003. Disciplinary Research Strategies for Assessment of Learning. *BioScience* 53: 1221–1228.

National Research Council. 2005. *How People Learn: Brain, Mind, Experience, and School.* Committee on Developments in the Science of Learning, J.D. Bransford et al, Editors. Division of Behavioral and Social Sciences and Education. Washington, DC: The National Academies Press.

National Research Council. 2005. *How Students Learn: Science in the Classroom.* Committee on *How People Learn*, A Targeted Report for Teachers, M.S. Donovan and J.D. Bransford, Editors. Division of Behavioral and Social Sciences and Education. Washington, DC: The National Academies Press.

Pryer, Kathleen M., Harald Schneider, Alan R. Smith, Cranfill, R., Paul G. Wolf, P.G. Wolf, and S.D. Sipes. 2001. Horsetails and Ferns Are a Monophyletic Group and the Closest Living Relatives to Seed Plants. *Nature* 409: 618–622.

Pryer, Kathleen M., Eric Schuettpelz, Paul G. Wolf, Harald Schneider, Alan R. Smith, and Raymond Cranfill. 2004. Phylogeny and Evolution of Ferns (Monilophytes) with a Focus on the Early Leptosporangiate Divergences. *American Journal of Botany* 91(10): 1582–1598.

Wilson, C.D., C.W. Anderson, M. Heidemann, J.E. Merrill, B.W. Merritt, G. Richmond, D.F. Silbey and J.M. Parker, 2006. Assessing Students' Ability to Trace Matter in Dynamic Systems in Cell Biology. *Life Sciences Education* 5: 323–331.

Wilson, R.W. and J.A. Mullins. 1980. Living or Non-living: A Game of Twenty Questions. *American Biology Teacher.* 42:566, 569.

CHAPTER 3

Crimmins, Shawn M., Solomon Z. Dobrowski, Jonathan A. Greenberg, John T. Abatzoglou, and Alison R. Mynsberge. 2011. Changes in Climatic Water Balance Drive Downhill Shifts in Plant Species' Optimum Elevations. *Science* 331: 324–327.

Dobzhansky, T. 1973. Nothing in Biology Makes Sense Except in the Light of Evolution. *American Biology Teacher* 35: 125–129.

Erschbamer, B., P. Unterluggauer, E. Winkler, and M. Mallaun. 2011. Changes in Plant Species Diversity Revealed by Long-term Monitoring on Mountain Summits in the Dolomites (northern Italy). *Preslia* 83: 387–401.

Hemingway, C.A. , W. Dahl, C. Haufler, and C. Stuessy. 2011. Building Botanical Literacy. *Science* 331: 1535–1536.

Pauli, H. et al. 2012. Recent Plant Diversity Changes on Europe's Mountain Summits. *Science* 336: 353–355.

Sperry, J. S. 2003. The Evolution of Water Transport and Xylem Structure. *International Journal of Plant Science* 164(3 Suppl.): S115–S127.

Credits

Photographs, line art, and tables are open sources unless otherwise indicated.
Title page © Simon Malcomber

CHAPTER 1

Chapter 1 title page © James Riser. Box 1 © G.E. Uno. Fig. 1.1 © M.D. Sundberg.
Fig. 1.2 From *Introduction to the Fine Structure of Plant Cells* by Myron C. Ledbetter and Keith R. Porter © 1970 With permission of Springer Science+Business
Media. Fig. 1.3–1.6 © M.D. Sundberg, except 1.15A (open acess). Fig. 1.7A © Clinton Steeds, Flicker CC-BY-SA-2. Fig. 1.7B © Jeffrey O. Gustafson, CC-BY-SA-3.0.
Fig. 1.8–1.10 © M.D. Sundberg. Fig. 1.12 © M.D. Sundberg. Fig. 1.13 © iStock. Fig.
1.14 From J. L. Harper 1970, originally published in J. L. Harper and J. Ogden.
1970. The reproductive strategy of higher plants I. The concept of strategy
with special reference to *Senecio vulgaris L. Journal of Ecology* 58(3): 681–689. ©
Wiley. Used with Permission. Fig. 1.15 From F.I. Woodward and C.K. Kelly. 1995.
The influence of CO_2 concentration on stomatal density. *New Phytologist* 131:
311–327. © 2006, John Wiley and Sons. Used with Permission. Fig. 1.16–1.21 ©
M.D. Sundberg. Fig. 1.22, 1.26 From *Basic Biology Course, Unit 2 Organisms and
their Environment, Book 4: Ecology* Game by M.A. Tribe and D. Peacock. © 1976
With permission of Cambridge University Press. 1.24 © M.D. Sundberg, except
insets (open access). 1.28 © G.E. Uno. Fig. 1.29 From L. Dayton © 2001. Plant
Physiology Philodendrons like it hot and heavy. *Science* 292: 186. Reprinted with
permission from AAAS. Image © Roger Seymour used with permission.

Table 1.1 © M.D. Sundberg. Table 1.2 Modified from U.S. Department of Interior, Bureau of Land Reclamation "Law of the River" (http://www.usbr.gov/lc/
region/g1000/lawofrvr.html) and U.S. Geological Service "Climatic Fluctuations, Drought, and Flow in the Colorado River Basin" (http://pubs.usgs.gov/
fs/2004/3062/). Table 1.3 From H. Dittmer. 1937. A qualitative study of the roots
and root hairs of a Winter Rye plant (*Secale cereale*). *American Journal of Botany*
24(7): 417–420. © The Botanical Society of America. Used with Permission. Table
1.4 From *Introduction to Botany Workbook* by Gordon Uno © 2010 Kendall Hunt.
Tables 1.5–1.10 From *Basic Biology Course, Unit 2 Organisms and their Environment, Book 4: Ecology* Game by M.A. Tribe and D. Peacock. © 1976 Used with permission of Cambridge University Press.

CHAPTER 2

Chapter 2 title page © Julia Nowak. Fig. 2.1–2.2 © M.D. Sundberg. Fig. 2.3
From *A Textbook of Botany for Colleges* by William Ganong 1917. The MacMillan

Company. Fig. 2.4–2.15 © M.D. Sundberg. Fig. 2.16A, 2.16B © Forest and Kim Starr, CC-BY-SA-3.0. Fig. 2.16C © Christian Fisher, CC-BY-SA-3.0. Fig. 2.16D © David Eickhoff, CC-BY-SA-2.0. Fig. 2.17, 2.19–2.20, 2.22–2.23 © M.D. Sundberg. Fig. 2.18A © S.T. Malcomber. Fig. 2.18B © Brian Thorson. Fig. 2.21A © Christian Fisher, CC-BY-SA-3.0. Fig. 2.21B © Adam Mickiewicz, CC-BY-SA-3.0. Fig. 2.24–2.25 © S.T. Malcomber. Fig. 2.26 From Pryer, Kathleen M., Eric Schuettpelz, Paul G. Wolf, Harald Schneider, Alan R. Smith, and Raymond Cranfill. 2004. Phylogeny and Evolution of Ferns (Monilophytes) with a Focus on the Early Leptosporangiate Divergences. *American Journal of Botany* 91(10): 1582–1598. © 2004 Botanical Society of America. Used with Permission. Fig. 2.27 From D'Avanzo, C. 2008. Biology Concept Inventories: Overview, Status, and Next Steps. *BioScience* 58: 1–7. © 2008 University of California Press. Used with Permission. Fig. 2.28 From Ebert-May, D., J. Batzli, and H. Lim. 2003. Disciplinary Research Strategies for Assessment of Learning. *BioScience* 53: 1221–1228. © 2003 University of California Press. Used with Permission. Fig. 2.29–2.31 © M.D. Sundberg.

CHAPTER 3

Chapter 3 title page © Simon Uribe-Convers. Fig. 3.1 From *Introduction to Pest Management*, 3rd Edition by Robert L. Metcalf and William H. Luckmann, editors. © 1994 John Wiley and Sons. Used with permission. Fig. 3.2,3.3, 3.5 From *Principals of Botany*, 1st Edition, by Gordon Uno, Richard Storey, and Randy Moore. © 2001 McGraw Hill. Used with permission. Fig. 3.4 from I.W. Bailey and W.W. Tupper. 1918. Size Variation in Tracheary Cells: I. A Comparison Between the Secondary Xylems of Vascular Cryptogams, Gymnosperms, and Angiosperms. Proceedings of the American Academy of Arts and Sciences Vol. 54. No. 2. Fig. 3.6 © C.A. Hemingway, except Saguaro © Ken Bosma, Magnolia © Beatriz Moisset. Fig. 3.8 From *Essay on the Geography of Plants* edited by S.T. Jackson. © 2010 University of Chicago Press. Used with Permission. Fig.3 9 Reprinted by permission from Macmillan Publishers Ltd: © 2012. Nature Education. From Irwin N. Forseth. 2012. Terrestrial Biomes. Nature Education Knowledge 3(1): 11. Fig. 3.10 USDA. Figs. 3.11,3.14 From Pauli, H. et al. 2012. Recent plant diversity changes on Europe's mountain summits. *Science* 336(6079): 353–355. © 2012. Reprinted with permission from AAAS. Fig. 3.12 From Pauli, H., Gottfried, M., Hohenwallner, D., Reiter, K., Casale, R. and Grabherr, G. (2004). The GLORIA field manual–multi-summit approach. European Commission, DG Research. Office for Publications of the European Community, Luxembourg. Fig. 3.13 From Photographic Documentation Vegetational Changes in Northern Great Plains. by Walter S. Phillips. 1963. Report 214, University of Arizona, Ag. Expt. Station. © 1963. University of Arizona. Reprinted with permission. Fig. 3.15 From B. Erschbamer, P. Unterluggauer, E. Winkler, and M. Mallaun.